HDTV
High-Definition Television

Stan Prentiss

TAB BOOKS Inc.

Blue Ridge Summit, PA

FIRST EDITION
FIRST PRINTING

Copyright © 1990 by TAB BOOKS Inc.
Printed in the United States of America

Library of Congress Cataloging in Publication Data

Prentiss, Stan.
HDTV : high-definition television / by Stan Prentiss.
p.   cm.

ISBN 0-8306-9272-X   ISBN 0-8306-3272-7 (pbk.)
1. High definition television.   I. Title.

TK6679.P74   1989   89-36604
621.388—dc20   CIP

TAB BOOKS Inc. offers software for
sale. For information and a catalog,
please contact TAB Software Department,
Blue Ridge Summit, PA 17294-0850.

Questions regarding the content of this book
should be addressed to:

Reader Inquiry Branch
TAB BOOKS Inc.
Blue Ridge Summit, PA 17294-0214

Acquisitions Editor: Roland S. Phelps
Production: Katherine Brown

Contents

Acknowledgments

Understanding that all written material submitted to the Federal Communications Commission and the Advanced Television Systems Committee is "in the public domain," I would, nonetheless, like to express my appreciation to those individuals and companies or corporations who contributed so much by their excellent technical presentations and highly innovative engineering efforts. In consequence, wherever their text or illustrations are used throughout the book, due credit is given without reservation.

Such recognition includes: David Sarnoff Research Laboratories; Zenith Electronics Corp.; North American Philips; Production Services, Inc.; NHK and BTA (Japan); Hitachi; Matsushita Electrical Industrial Co., Ltd.; High Resolution Sciences; Osborne Associates; Scientific Atlanta; Digideck; Dolby Signal Processing and Noise Systems; dbx, Inc.; Quanticon; The Advanced Television Research Program, MIT; The Del Rey Group; New York Institute of Technology; and those individuals among the FCC Advisory Committee, the Advanced Television Systems Committee, the Advanced Television Test Center, and the Working Party Chairmen who made all this possible.

Special appreciation is further extended to Advisory Chairman Richard Wiley; William Hassinger, FCC; Robert Hansen, Zenith; Dr. James Carnes, Sarnoff Labs; Edmund Williams, Ben Crutchfield, and Charles Rhodes, ATTC; Birney Dayton, Grass Valley Group; Lawrence Thorpe, Sony; Dr. D.J. Donahue, Thompson Electronics; Robert Oblack and Bill Benedict, Tektronix; with special mention for Elizabeth Sadove, Telecommunications Policy Analyst with Chairman John Markey's House Telecommunications and Finance Subcommittee, U.S. House of Representatives.

Introduction

This book is a definitive discussion and report on U.S., European, and Japanese developments in extended (EDTV) and high definition television (HDTV), as well as examples of the latest improved television (IDTV) immediately available in stores around the country. IDTV simply takes advantage of recently maturing digital technology that removes most artifacts such as cross color and luminance problems from the picture. IDTV sometimes doubles horizontal scan in what is currently known as 1:1 progressive scan versus 2:1 conventional dual-field interlace, delivers precise microprocessor/microcomputer picture and stereo sound control, and agreeably digitizes audio for almost remarkable improvement over ordinary BTSC-dbx multichannel (MTS) sound.

Once commercial enterprises and consumers have become used to these obvious advances, EDTV can then extend picture detail, often in 5:3 aspect ratios, rather than our standard 4:3 horizontal and vertical measurements of today. Progressive scanning for smooth, uninterlaced images should continue, fully digitized sound could be improved, and probably an additional VHF or UHF channel will be needed to accommodate not only further digital requirements, but also expanded aspect ratios to even 16:9. In the meantime, there will certainly be Japanese, possibly European, and perhaps even U.S. recorders and receivers with extended luminance bandwidths, in addition to common bus input ports to accommodate Y (luminance) and C (chroma) as well as RGB (red, blue, and green), video/audio, and teletext, plus the usual CATV and UHF/VHF broadcast channels nationally available.

HDTV should rear its delightful head in possibly two forms: *Simulcast*, where one pure NTSC and another HDTV high definition channel reaches all existing receivers; or in *augmented* structure where the second channel occupies half or a full bandwidth of another channel so that maximum bandwidth, expanded aspect ratios, and digitized audio in two or four channels becomes fully available. Having already experienced the forerunners of all of these projected systems, we can happily promise everyone is on the threshold of unique commercial possibilities and home entertainment that's unrivalled since the advent of broadcast television.

These anticipated developments can't occur all at once since many ongoing electronic designs must be developed, tested, and proven, and the Federal Communications Commission has yet to deliver its ultimate blessing, but progress is coming rapidly, and certain breakthroughs could expedite some systems considerably faster than many might think. Meanwhile, over-the-air terrestrial and satellite transponder testing has already begun, augmenting laboratory Advanced Television Test Center (ATTC) activity making progress as the newest systems are received.

In this EDTV/HDTV endeavor, your author has been an active member of three working parties under the Advanced Television Systems Committee, and hopes that this report will not only offer an encyclopedia of useful facts, but sufficiently interpreted information from industry, the Federal government, and the ATSC to aid in forming a substantive judgment on the merits of the majority of proposed high(er) definition television systems.

1

High Definition Television: Who, What, Why, When, and How

AS IN THE FIVE REQUISITES OF NEWS GATHERING, HIGH DEFINITION TELEVISION (HDTV) is finding its way into public, business, and engineering consciousness in print, on the air, and by the spoken word. Industry, advisory, planning, and implementation groups have all been very hard at work attempting to analyze and evaluate the various systems proposed from Europe, Japan, and this country, and are coordinating the effort with our northern neighbor Canada—all within the general guidelines laid down by the Federal Communications Commission in MM Docket 87-268 of September 1, 1988.

Defining the on-going development as ATV, or advanced television techniques, the Commission states that it includes either/or Enhanced Definition Television (EDTV) or High Definition Television (HDTV). Further, broadcasters must "tentatively" use only TV spectrum now allocated, and existing NTSC services for the foreseeable future and during any transition period. Translated, this means no disruptive changes in delivery or reception of common television signals while development of new systems continue, and probably far into the future, since public reaction would be more than considerable if 140,000,000 TV sets were outmoded all at once. In other "services" such as cable and satellite television, or non-broadcast media, the FCC will not *retard* independent introduction of ATV, but remains "sensitive" to compatibility of various delivery methods. The 9 MHz FM Japanese MUSE system, for instance, would not be approved for terrestrial broadcasts since it is outside the 6 MHz standard NTSC system.

Conversely, MUSE may well be delivered sporadically or otherwise over satellite signals or even cable, both of which have enough capacity to accept either full or compressed MUSE since satellite transponders have at least a 36

MHz bandwidth, and fiberoptic cable can accommodate 50 MHz. For FM modulation, sideband pairs in the Japanese system would occupy 27 MHz under existing parameters or current specifications. Ordinary AM cable carriers, of course, would be considerably less. For now, at any rate, the ATVers will have to stay within the compatible 6 MHz broadcast band if they want their products on terrestrial air.

This does not mean, however, that the FCC would refuse approval of a second broadcast channel for HDTV sometime in the future. But in any event, the cost of a broadcast station changeover is estimated between $1 to $4 million to accommodate whatever system is approved. There also seems little chance at the moment that UHF (ultra high frequency) spectrum *above* 806 MHz could or would be set aside for second channel HDTV broadband operation. Land mobile and a few remaining active repeater stations apparently have a lock on these frequencies barring a radical change in thinking. Nonetheless, reregulation is always possible under different political circumstances and national direction. Congress, too, may well become quite sensitive to the issue should cable TV, VCRs, and satellite delivery systems threaten over-the-air broadcasting. Political persuasion and campaign contributions are powerful motivators every two years in the House of Representatives and four years in the Senate. In an election year, almost anything is possible, especially for big buck interests.

This does not mean that active FCC studies of "available spectrum" aren't proceeding in the meantime. The Spectrum Engineering Division of the Office of Engineering and Technology has already issued an interim report on the *"Estimate of Availability of Spectrum for Advanced Television in the Existing Terrestrial Broadcast Bands,"* and will probably follow this with additional studies as new data is requested and suggested. For the time being, such studies are considering the implications of ATV not requiring additional spectrum, or those which need 3- or 6-MHz above and beyond the conventional 6 MHz within NTSC. Both adjacent channels, as well as others in TV normally-assigned frequencies are possibilities. Of course, any such channel separation for either EDTV or HDTV would seemingly require two receiver tuners and added analog/digital subsystems for adequate reception and display. As we advance further into the technical aspects of this vastly important engineering accomplishment, further references will be made to this report prepared by Bob Eckert, Julius Knapp, and Ray LaForge, dated August 1988.

PREPARATIONS

NBC and the David Sarnoff Research Laboratories, North American Philips, Japan's MUSE, Zenith, The Del Rey Group, Faroudja Laboratories, and many others are working feverishly to perfect systems that are fully acceptable to the video industry as well as the FCC. At the same time, some are still complaining that the possible international standard of 1125 SCANNING LINES and

60 Hz fields/second will make trouble for our NTSC system of 525 lines and 59.94 Hz now in use. But the U.S., Canada, and Japan are considering this standard and we foresee no change, unless some unforeseen development occurs that upsets all system projections thus far. Anything is possible but somewhat improbable, although some manufacturers are still advocating production standards that follow adoption of purely national transmission systems.

As the various system proponents grouse about this and that but continue their own particular developments, Southern Bell and others of the Bell phone system are demonstrating that HDTV is immediately available to almost anyone via multipoint switching, fiberoptic cable, and Japanese cameras and receivers. If all this is possible now, what's to prevent the telephone companies from putting video directly into anyone's home? Should the FCC approve, the telephone companies would undoubtedly move in this direction immediately. Such a prospect obviously concerns the cable companies and broadcasters considerably. But for now, fiberoptics cable isn't everywhere, only experimentally in 256 homes in Heathrow, Forida.

Many factors are entering the high definition picture that further compound even existing problems. Eventually, however, we will probably find there's room for all in the communications business since it has grown so enormously large. And with keen competition everywhere, some pretty good bargains should become available once the initial systems are in place and operating. But such ''bargains'' will only be relative, since extended or high definition equipment is certain to be more expensive than even high-end receivers of today. A special savings account of a couple of thousand dollars might help.

In the meantime, while HDTV climbs out of rompers, Yves Faroudja has ''discovered'' Super NTSC, which amounts to a prefiltering encoder, Hitachi's Dr. Fukinuki plots NTSC in vertical and horizontal time dimensions and discovers a ''hole'' or empty space next to the color subcarrier. This he calls FUNCE, for fully compatible EDTV, since he fills the hole with an additional subcarrier for extra bandwidth. Thereafter, Matsushita (Panasonic) finds that suppressed subcarrier modulation for the I and Q color information will permit addition of another signal allowing the usual NTSC 4:3 aspect ratio to be expanded to the HDTV-targeted 16:9 ratio for broad-screen viewing. All of which, obviously, is still within the traditional 6 MHz bandwidth of our existing system. Then there are also special effects available in some of Philips' receivers for removal of artifacts and better chroma/luma separation, plus double line scan, reaching the consumer market in 1989. So some of the impending advantages of improved, extended, or high definition television are already here, and could make their appearance in the near or fairly near future if broadcasters and manufacturers are willing to spend an extra nickel for interim advantage. Nonetheless, all of the foregoing indicates a revolution in picture quality that's really the first since the introduction of the comb filter by Magnavox (now North

American Philips) in the 1970s. Rome, manifestly, could never "have been built in a day," not even electronically. The other stickler in these ambitious assumptions assumes complete compatibility with NTSC and no interference—a stipulation that may not be totally acceptable to old or cheap diode detector receivers. But then, again, this may not become an overwhelming consideration. Unlike the wagon, the automobile was not required to have steel-tired wheels.

THE OUTLOOK

While Western Europe and Japan have government support and cooperation in their endeavors, the U.S. depends primarily on private industry to fund the various projects, do engineering, provide system testing, and pick a recommended system or systems—thereby delaying introduction of an enhanced or high definition system by one to several years. While Europe and Japan will probably be on satellite transponders by 1991 at the latest, U.S. broadcast systems may not be on the air until 1995. This isn't all bad, because we can obviously profit by European and Japanese mistakes and shortcomings in the meantime. But the advent of Japanese VCRs and HDTV monitors will certainly put more pressure on U.S. companies to produce equal or better consumer products as rapidly as possible. In addition, it may offer a little breathing room for hard-pressed designers to prove their products before rushing into precipitous production.

But in the meantime, do we once again succumb to Japanese consumer penetration equivalent to that experienced in the past 10-15 years? Perhaps this time some of our own—or at least American-made video components will win out. There are some indicative stirrings among certain manufacturers that lend hope, even though most activities involve transmissions. Among receiver manufacturers only Zenith is still American owned, and this may not last forever if consumer products continue to make this electronics giant lose money. We are aware, however, that probably Thompson/RCA, North American Philips, and Zenith are capable of producing adequate receivers when the time comes. In addition, Zenith has also introduced an entirely different whole-channel transmission means of its own that is even now quite interesting to the FCC and others, if the hardware operates as well as projected. This and the other current 15-odd system contenders will be discussed generally in the next chapter.

System recommendations are sure to result from investigations by the Advanced Television Test Center located in the Washington, D.C. metropolitan area, whose chief scientist/engineer is Charles W. Rhodes, formerly of Philips, Scientific Atlanta, and Tektronix. Over $3.5 million has been set aside by the National Association of Broadcasters (NAB), Association of Maximum Service Telecasters (MST), Television Operators Caucus (TOC), TV networks NBC, CBS, and ABC, in addition to the Association of Independent Telecasters (INTV), and Public Broadcasting. Testing and retesting may last for as much as two years, depending on changes, additions, and any protested results. With so many proponents involved, we really can't be sure of any precise time limit. A detailed review of system tests and criteria can be found in chapters 6 and 7.

THE EUROPEAN SCENE

Thanks to Director Richard Kirby of the International Radio Consultative Committee (CCIR), we do have a progress report on HDTV directly from Geneva, Switzerland, outlining European considerations and developments in late 1988. It notes that the CCIR has been studying high definition television since 1972, reviews HDTV developments between then and 1985, and includes an analysis of current conditions by study group 11.

Research on high definition television continues under Eureka industry, educational, and other institutions established in 1985 and 1986. It corresponds to the U.S. Advanced Television Systems Committee (ATSC), founded in 1982, and the Canadian Advanced Broadcast Systems Committee (CABSC), working directly with U.S. interests in developing a broad North American HDTV standard.

Study Group 11 has now drafted "*A Global Approach to High Definition Television*" and delivered its findings to both ATSC and CABSC for their consideration. Meanwhile, the Japanese have adopted their own HDTV studio standard prepared by the Broadcast Technology Association (BTA), a report now being analyzed by both Europe and the Americas.

Surprisingly, the CCIR study group recommends similar, if not identical parameters established previously by Canadian-Japanese interests of:

1125 lines/frame
1035 active lines/frame
2:1 interlace
16:9 aspect ratio
609 Hz fields/sec
33.750 kHz line frequency

and that chromaticity coordinates for RGB primary colors be set at:

	X	Y
R	0.630	0.340
G	0.310	0.595
B	0.155	0.070

Although these coordinates do not now correspond to those established long ago by the Federal Communications Commission, many Europeans still believe that choosing a "wider color gamut" would be beneficial. This could be a sticking point for any worldwide system if engineering recommendations remain rigid, or we decide our own parameters are already broad.

Europeans would also set their reference black, white, and sync levels at 0.700, and ± 300 mV, respectively, and nominal bandwidths at 30 MHz. In digital structure, luminance sampling would occur at 74.25 MHz and color difference signals at 37.125 MHz, with quantized PCM coding at N bits/sample. But analog-to-digital timing relationships remain a subject for study in 1989 and 1990, as does codeword usage.

In analog synchronizing waveforms, the sync signal would become tri-level bipolar, the normal line blanking interval 3.77 μsec, field blanking at 45 H, and the reference clock frequency remaining at 74.25 MHz nominal. Figure 1-1 supplies sync timing and line/field blanking details as originally proposed, and is printed without comment.

As for HDTV conversion to both European and NTSC standards, the report states that a system converter from 1125/60 to 625/50 and 525/60 has already been developed by NHK (Nippon Hoso Kyoka) which delivers "satisfactory results." The mean signal "impairment," meaning signal degradation, is apparently no more than 0.5 to 0.7 of selected scenes.

As for HDTV to film conversions, the Society of Motion Picture and Television Engineers (SMPTE) are said to have concluded an increase of frame rate from 24 to 30 frames per second "significantly improves the quality of the projected images by virtue of permitting higher screen brightness without full field flicker and by increasing dynamic resolution." But combinations of film and HDTV could produce temporal difficulties without very special precautions, according to CCIR. In summary, Study Group 11 believes that conversions between European and U.S. HDTV would be possible, even though the converter would probably be "complex and sophisticated."

The Committee, nonetheless, does report that HDTV studio equipment is at an advanced stage by over 20 companies, and that converters from the announced 1125-line, 60 Hz standard have been successful in 24-30 frame rate difference.

Nonetheless, many European countries still want to cling to their current field rate, along with progressive scanning, and the CCIR is asked to recommend the following:

1152 active scanning lines
50 Hz field rate
Progressive scanning
16:9 aspect ratio
1920 luminance samples/active line, and 960 for color difference.

But there remains some support for a 60 Hz version and efforts will continue in the hopes of reaching a workable compromise. Otherwise there can be no universal standard.

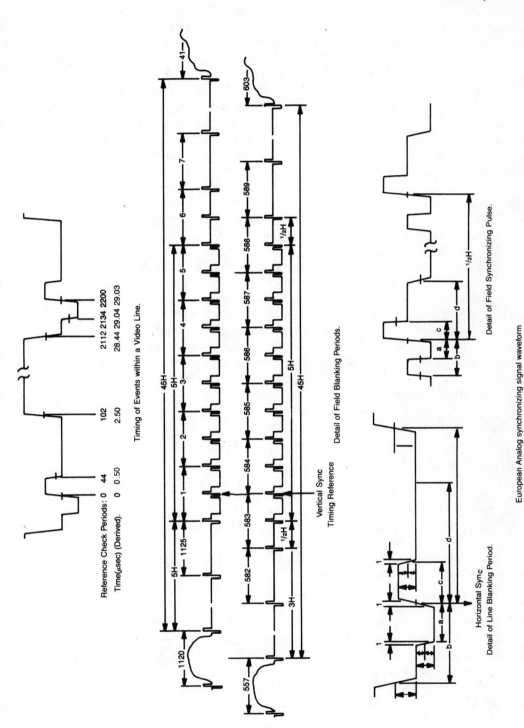

Timing of Events within a Video Line.

Reference Check Periods:	0	44	102	2112	2134	2200
Time(µsec) (Derived).	0	0.50	2.50	28.44	29.04	29.03

Detail of Field Blanking Periods.

Vertical Sync
Timing Reference

Detail of Field Synchronizing Pulse.

European Analog synchronizing signal waveform

Horizontal Sync
Detail of Line Blanking Period.

Fig. 1-1. The European analog synchronizing waveforms and special details.

As studies and arguments in every direction progress, HDTV demonstrations have already taken place at the Berlin International and Brighton, England, some of which have included both D and D2-MAC—multiplexed analog component signals in two versions—plus a digitally assisted television technique to pass motion-adaptive control information to the decoder. Fiberoptics cable transmissions are not reported in any detail, although some video has already been demonstrated at a gigabit rate. There is considerable talk, however, of relatively reasonable costs for home receivers operating in systems with considerable bandwidth compression that would also include standard, conventional reception. But for now, European manufacturers see the cost of dual standard integrated circuits as prohibitive and would delay a D2-MAC system. For satellite transmissions, some mention was made in the reports of working with Japan's MUSE system, with subsequent improvement of frequency processing between dc and 4 MHz that seems to avoid interframe aliasing.

Sound channels in the MAC format are being investigated also, and include HD-MAC, D2-MAC, and C/D-MAC—all, of course being multiplexed. D2-MAC seems to have a multiplexing capacity of 1.5 megabits, corresponding to some four 15 kHz maximum frequency channels. In C/D-MAC the useful bits per line are doubled over D2-MAC, permitting eight good quality channels or 16 medium quality ones. There's also a possibility, according to the CCIR report, that European HDTV might be delivered terrestrially via amplitude modulation over regular VHF and UHF channels. And should HDTV TV signals match those of MAC at baseband, there could be compatibility between the two services. A single channel, however, would require a bandwidth of 14 MHz.

HDTV World Standard Is Nearer

A semi-final meeting of the International Radio Consultative Committee's Study Group 11 has taken place in Geneva, and the U.S. delegation became an overall winner.

Our Advanced Television Systems Committee gained approval for colorimetry and transfer characteristics, based on the findings of international colorimetry experts the previous week in Australia. According to ATSC chairman James McKinney, ''never before have all countries been able to agree on such basic definitions of colors and other values which would assure that all television viewers will see exactly the same video pictures regardless of where they are watching television.''

The only outstanding points remaining are the common image format and the 50/60 Hz questions. The U.S. supports 1080 scanning lines and 59.94 or 60 fields/second while the Europeans want 1152 lines and 50 fps.

A final meeting of the CCIR on these subjects will take place during March 1990 before the National Association of Braodcasters convention in Atlanta, Ga. Neither question may be resolved since most all of Europe operates on 50 Hz and some on 240 volts. But a standard now could easily be set for line and field conversion should neither side compromise its position. Eventually, of course, digital television will take over and then a common bit rate could much more easily be decided upon—but this is years away, probably in the next century.

JAPAN WITH ANOTHER FIRST

Never let it be said the sons of Nippon are bashful. In consumer products they're often state of the art, show partial leadership in solid state, and now own considerable U.S. real estate. At the moment they are a leading element in the development of U.S.-Japanese high definition television since they already have cameras, VCRs, monitors, and other equipment that obviously produces HDTV both here and in Japan. It's a lead that will be difficult to overtake, regardless of the U.S. system approved. Fortunately, their original MUSE (for Multiple Sub-Nyquist Encoding) is an FM system of very wide bandwidth that has already been rejected for terrestrial broadcasts in the U.S. by the FCC. But MUSE reputedly has another five or six versions, the latest being a vestigial sideband amplitude modulated (AM) system of two channels based on the 1125-line, 60 fields. Baseband video is said to be 8 MHz. This we'll investigate much more thoroughly in the next chapter detailing the various systems. In a recent HDTV committee meeting in Washington, Japanese MUSE representatives facetiously suggested on-site demonstrations in Tokyo. One of the other committee members, equally waggishly, wanted to know if the Japanese would buy a Boeing 747 "to take us there?"

In the interim, while our committees study and the FCC examines and waits, NHK has opened and closed the 1988 Olympics in Seoul with satellite broadcasts of the ceremonies, but regular taping of the athletic proceedings have been provided by AMPEX to the Koreans with their NTSC one-inch VPR-6 high-reliability and excellent slow and stop motion abilities. A total of more than 50 of these machines were involved in the effort, with master control using 18.

The Japanese HDTV effort is not new. In 1970, NHK began its opening research, sharing results with Japan's 11 TV manufacturers as progress continued. The government then began coordinating these activities and opening the way for future public acceptance. Already one or more channels have been set aside on a forthcoming communications satellite due for launching by 1991 in anticipation of a $30 billion market. Meanwhile, public demonstrations at various locations are continuing to maintain the excitement, and Japanese salesmen and engineers are trying hard to sell MUSE in one or more variations to Europe and the U.S.

Will this wider effort succeed? Europe is quite nationalistic and many there are convinced that if MUSE is accepted it will mean the end of their own HDTV industry. Similarly, the U.S. is concentrating on terrestrial HDTV service, with other considerations somewhat secondary. To Americans, it's obvious CATV and satellites can rather easily carry the wider bandpasses, especially while using fiberoptics cable. But if a single or two-stage system can be developed that's compatible with existing NTSC, all receiver owners can receive or reject HDTV, depending on pocketbooks and enthusiasm. At the moment, that's what committee meetings and testing is all about. We obviously subscribe somewhat to the 1125/60 Japanese/Canadian format, but how to do this and in how many stages and versions will consume considerable midnight oil before any concrete decision emerges. Our personal hope is that the deregulating FCC won't throw the problem back to industry for a destructive fight without selecting a prime system. The recent divisive memory of AM stereo is all too vivid as a vacillating disaster. In that case, the FCC issued a no-selection notice on March 4, 1982, and only recently has Motorola's C-QUAM become the apparent winner in both the U.S. and abroad. Lawyers opinions versus engineers analyses probably contributed heavily to that fateful day. We trust both the FCC and Congress are somewhat more sensitive and resolute tomorrow. It's distressing to see legal eagles attempt to fly on one wing.

Since we intend to detail the MUSE system rather exhaustively in the next chapter, only general characteristics of the Japanese method will be printed here in Fig. 1-2. The figure is taken from examples presented to the 1987 International Colloquium on HDTV held in Ottawa, Canada. Following the Colloquium, MUSE-E and a 27 MHz transponder on ANIK-C demonstrated HDTV broadcasts to audiences in three Canadian and four U.S. sites, introducing some 100,000 individuals to the new techniques.

System Description	Motion-compensated multiple Subsampling System (multiplexing of Y and C Signals is Done in TCI format
Scanning Rate	1125 Lines/60 Fields/2:1 Interlace
Bandwidth of Transmitting Baseband Signal	8.1 MHz
Sampling Clock Rate	16.2 MHz
Reproduced Signal Bandwidth — Y Signal	22 MHz (for Stationary Portions of Picture) 14 MHz (for Moving Portions of Picture)*
Reproduced Signal Bandwidth — C Signals	7.0 MHz (for Stationary Portions of Picture) 3.5 MHz (for Moving Portions of Picture)*
Synchronizing Signal	Positive Polarity with Respect to Video Signal Polarity

*These values should be 16 MHz for Y and 4 MHz for C respectively, if a perfect digital two-dimensional filter could be used.

Fig. 1-2. Basic video specifications attributed to the MUSE system.

Whether the general public on the several continents are willing to spend several thousand dollars for an improved video system with upgraded audio probably depends on a number of factors, including image detail, cost, evident advantages, and programming. This will be far from the quantum leap between radio and television. Before TV, there was little more than flickering motion pictures and good old radio. HDTV will probably have to be sold in a number of ways for the public to substantially open their pocketbooks, since most don't yet seem to appreciate the 25% better resolution, definition, and color provided by today's effective comb filters.

For now, Japan's Sony, Hitachi, Matsushita, and Toshiba are investing many millions of yen in HDTV, continually increasing inputs of both money and personnel, with a great deal of the effort directed toward gaining a solid foothold in both Europe and the U.S. Should they fail, the result might become a consumer electronics catastrophe; but if they win, Anglo-American HDTV and some allied industries may have to close up shop!

Yes, the Japanese do have a working HDTV system, and yes, this system can work on cable and satellites. But what Europe and our own FCC decide, could well determine the future course of all television on at least our two continents, and possibly the world. This is a real test of innovative engineering and legislative guts! By late 1991–1992 we may know the winner, regardless of whether the end product originates in the Union, the U.K., or the Land of the Rising Sun. There's just one consolation, if selection of HDTV follows the pattern of multichannel TV sound (BTSC-dbx), then it's probable a U.S. manufacturer will take the cake, despite some generally agreed-upon international standard—if there is one. There are already some very solid and innovative proposals submitted to evaluation and testing committees that seem to have an excellent chance. Fortunately, American technology is far from dormant, just very profit conscious and highly competitive.

THE AMERICAN EFFORT

In his testimony before the House Telecommunications and Finance Subcommittee on Advanced Television Technologies, FCC Advisory Committee Chairman Richard Wiley declared that "in my judgment, HDTV represents potentially the most significant advance in the state of the television art since the advent of color" (December 17, 1953). He continued that "on the basis of very preliminary engineering studies, the Advisory Committee believes . . . there may be sufficient spectrum capacity in the current TV allocations to permit all existing stations to provide advanced television service through either an augmentation or simulcast approach." (This means standard signals on one channel and an augmented or HDTV on a second VHF or UHF channel that would occupy either an additional 3 or 6 MHz.)

Former FCC Chairman Wiley also said his committee did not "foreclose" possibilities of enhanced definition television (EDTV) as an evolutionary step

TECHNOLOGY OR ATTRIBUTE	AVELEX-HDTV	EDTV-I	EDTV-II	SuperNTSC	HRS-CCF	HD-NTSC	RC	CC	MUSE-6	MUSE-9	NARROW MUSE	MUSE	VISTA	OCS	HDMAC-60	HDNTSC 6+3	HDNTSC 6+6	GENESYS	QuanTV	ACTV-E	ACTV-I	ACTV-II	HDB-MAC	SC-HDTV
PROPONENT	AVELEX	BTA	BTA	FAROUDJA	HI RES SCI	IREDALE	MIT	MIT	NHK	NHK	NHK	NHK	NYIT (GLENN)	OSBORNE	N. A. PHILLIPS	N. A. PHILLIPS	N. A. PHILLIPS	PSI	QUANTICON	SARNOFF	SARNOFF	SARNOFF	SCI. ATLANTA	ZENITH
PRESENT AT 11/14/88 MEETING		•		•	•	•	•	•	•	•	•	•	•		•			•	•	•	•	•	•	•
RECEIVER COMPATIBLE	•	•		•	•	•	•		•	•			•		•			•	•	•	•	•	•	•
3 MHz AUGMENTATION CHANNEL										•			•		•			•			•			
6 MHz AUGMENTATION CHANNEL													•	•				•			•			
SIMULCAST									•			•							•					•
NON-SYSTEM (TECHNOLOGY ONLY)				•															•					
NON-BROADCAST (SATELLITE)												•			•								•	
LETTERBOX							•	•	•		•	•												
SIDE PANEL	•																				•		•	
SIDE PANEL STITCHING																					•	•		
PAN-AND-SCAN													•								•	•		
INTERLACED								•		•	•	•	•	•						•		•		
PROGRESSIVE SCAN	•	•					•	•							•	•					•	•	•	•
BEZEL CONCEALMENT							•													•	•	•		
INFO. IN LETTERBOX BARS							•	•		•	•													
VISUAL CARRIER QUADRATURE																				•	•	•		
FUKINUKI CHANNEL																				•	•	•		
SUB-BAND FILTERS (QMF)						•	•						•							•	•		•	•
LINE DIFFERENCING																				•	•			
MULTIFIELD SUBSAMPLING	•				•			•		•										•	•	•		
FREQUENCY INTERLEAVING											•	•												
QUADRATURE AM								•									•							•
SYNC REMOVAL								•					•	•		•	•							•
NONLINEAR PRE-EMPHASIS		•							•	•														•
COMPANDING.							•	•																•
TIME DISPERSION																								•
DIGITAL LOWS								•																•
ELEMENT SCRAMBLING							•	•																
ADAPTIVE MODULATION								•																

Fig. 1-3. Assessment of initial HDTV proponents by WP1.

toward full HDTV. He strongly opposed any reallocation of currently authorized UHF spectrum for other uses (such as HDTV). Too, he favored elimination of UHF channel separation requirements (called taboos), along with added interference protection "substantially less" now required by NTSC regulations.

On the other hand, Mr. Wiley observed that the public has already made a "very significant investment (perhaps $100 billion) in current NTSC receivers and technical video improvements and shouldn't obsolete this considerable outlay. Nonetheless, the new HDTV service," he continued, "could represent a 'tremendous opportunity' for American companies and workers—perhaps $40 billion a year." Considering the potential, Mr. Wiley warned "that we should not be satisfied with simply being a passive recipient of some other country's HDTV format, programming, or video receivers and associated equipment." And although Chairman Wiley said he was skeptical of the U.S. again becoming a major factor in television set manufacturing, he suggested that associated equipment, components, and set assembly could result in many American jobs, technology, and income.

Prior to his congressional testimony, FCC Advisory Committee Chairman Wiley submitted a detailed report to the Federal Communications Commission covering much the same information but in greater technical depth, especially enhanced definition television (EDTV). Most proponents, he said, would offer wide screen displays with possible aspect ratios of 53:3 instead of the usual NTSC 4:3, and could also feature digital sound. But many "would vary greatly" in the amount of information transmitted and in spectrum requirements: some would need one channel, others 1.5, and the third category, two channels, all of which would cover spectrums of 6 MHz, 9 MHz, and 12 MHz, respectively. See Fig. 1-3.

Those who advocate two channels suggest an existing NTSC compatible channel plus an "augmentation" channel, or one NTSC channel and a simulcast channel with an "incompatible" signal. Based on known compression bandwidth engineering, he added, full HDTV would need more spectrum than 6 MHz. However, a footnote in the report suggests that EDTV may offer such picture improvement and audio quality that the public could accept it as an HDTV alternative. We might add that any preference would depend considerably on both color/luma quality as well as cost, and changeover problems within the broadcast stations, largely dependent on the same prime factors. We'll probably have an inkling of this after some of the improved definition (IDTV) receivers reach the market in the 1989-1990 model year. If public fancy accepts these and wants more, then EDTV will similarly fly with gusto. And all this may be precisely the key indicating how HDTV will eventually go, as well as whose system triumphs. Several system proponents are already analyzing the same considerations. Mr. Wiley adds that the availability "of such an opportunity may be essential if broadcasters are to continue to serve as a viable means of providing important local service benefits to the American people." In other words, broadcasters will have to climb on the bandwagon pronto! We'll see . . ..

Chairman Wiley also observes that adjacent channel and co-channel receiver specifications will have to be tightened to guard against possible interference; therefore, "high priority should be assigned to continuing spectrum analysis." In the meantime, offerers should further develop their systems to prevent undue interference from these two problems.

As EDTV/HDTV development continues, the Advisory Committee Chairman suggested that work and production of advanced television systems should go forward regardless of progress on terrestrial broadcasting. In this way, he said, "every video medium would be free to develop enhanced forms of delivery."

Because of the perceived importance of HDTV, special advisory groups to the Planning Subcommittee have now been appointed: one to study "creative issues," and the other to investigate "consumer/trade" issues. And so the diverse opportunities open to U.S. industry, such as system development, receiver manufacturing, assembly, components, and so forth, are to be evaluated in relation to both the public and private sectors. Some 175 individuals are involved.

The main Planning Committee has already "examined the attributes" of advanced TV systems and adopted evaluation tests, spectrum spreads, electronic media interfaces, economic factors, and analyzed subjective consumer responses to the proposed systems. This effort will unfold a "blueprint" for future committee activities plus an overall analysis of ATV systems in both public and private sectors.

Test systems are to advance in three stages: propagation to study effects of transmit signal paths; laboratory investigations of comparison system performances; and over-the-air checks in VHF/UHF, and 2.5 GHz and 12 GHz ranges. These latter tests, of course, have already begun, and most should be complete by the time of book publication. If substantial results are available they will be included, probably in a separate chapter. The various system proponents are and have been required to deliver encoding and decoding equipment, a suitable transmit format, and interface with an HDTV display.

As for program material, the Committee suggests 35mm film and video cassette recorders. We would also add satellite broadcasts from studios and possibly CATV where systems have the necessary bandwidth. There are also video discs that might be used for the purpose. Receivers will probably have both RGB entry as well as chroma/luma components, plus the usual 75-ohm input tuners. And the Committee suggests that with manufacturing changes, productivity, and a low dollar exchange rate, there may be a "second chance" for domestic ATV receivers.

Proponent test procedures will be based on CCIR studies and recommendations, and subjective assessments are to follow objective hardware laboratory testing. Motion, cropping, various formats, viewing ratios, luminance, and pair comparisons are all part of the subjective evaluations.

The Creative Issues Advisory Group wants highest quality TV images and sound quality equal to compact discs. The objectives are pictures equal to 35mm film, with a ''backward compatible'' ATV system that can handle NTSC, but with enough ''headroom'' to adapt to further advances in technology. This Committee would like the FCC to ''move expeditiously'' but not so fast that major technical issues aren't settled. Prime tasks now are the completion of spectrum options and the development of an assignment plan—both of which should be attainable in 1990. Meanwhile, most work overseen by the Advisory Committee will pass to the System and Implementation Subcommittees, with the Planning Subcommittee involved in spectrum studies.

STEREO (3-D) TELEVISION

Yes, it did occur at halftime in Super Bowl XXIII on January 22, 1989, but no, it's not yet wholly suitable for either NTSC or HDTV—at least in this version and under all conditions.

Developed by Terry Beard, founder and president of Nuoptix, the football showing consisted of a 45-second commercial with carefully controlled motion, an ordinary television camera and lens, and a pair of paper and plastic glasses supplied by Coca-Cola to advertise Diet Coke. In relatively slow motion, the right eye sees through plastic with a magenta tint, and the left eye looks through another piece of plastic having a green tint. Time constants through the plastic are different, therefore the right eye does not perceive color motion as quickly as the left eye, and consequently our brain in matrixing the two creates the sensation of a three dimensional image.

Unfortunately, stationary or fast-moving scenes can't be perceived equally, and so this application is considerably limited. The glasses—reportedly 20 million of them—were said to have been originally produced for a TV show called ''Moonlighting,'' but were never used due to a prolonged writers strike. Viewers without the special glasses won't notice any difference in their pictures, since the usual NTSC transmission isn't changed.

In years past there have been other, more sophisticated, 3-D color transmissions but, for one reason or another, these have never been commercially developed and put into production. One problem has been secondary image delay (ghosting) that plays havoc with clean 3-D reception. Other reasons could be cost or other competing factors such as multichannel sound—and now HDTV.

Being on a full NBC network for the first time, however, may generate some further impetus towards additional development, but my hunch is that HDTV comes first with all its enormous market implications. Later—and this may take years—3-D may surface once more and be adopted if all problems are successfully solved. Meanwhile, we'll have to suffer along with doubled definition and resolution, plus digital sound. Tough life!

Perhaps Coke and Pepsico, Inc. might like to join hands and sponsor a year or so of research on this project since they dominate the $43 billion American soft drink industry. A couple of million in the pot might help.

While the Super Bowl version of 3-D TV didn't arouse any perceivable demand, it did sell plenty of Diet Coke and caused some people to think a little bit about HDTV. In sports, though, we understand that camera lens weights, extra lighting, expensive cameras, and broadband transmission media are all going to be technical and money problems, especially in football. As you may have already surmised, action scenes will probably convert to passbands approaching NTSC, while still shots can use HDTV to singular advantage.

There's a lot we don't know about HDTV just now, but network crews out in the field, and especially in sporting events, are learning more than a little. We may find that the new medium is super for studio work, but needs more study, work, and applications outside. All these quirks require just a little more time to acclimate and deliver a professional product.

2
FCC Guidelines

ALTHOUGH THE FEDERAL COMMUNICATIONS COMMISSION HAS MADE A TENTATIVE HDTV directional decision and instituted a "further notice of inquiry," there is still much to be done before that body makes its final reckoning of what shall and shall not be. It was declared "this could be a new golden TV age if the proverbial goose doesn't lay a sandstone egg." A more serious individual predicted that HDTV would penetrate 1% of the TV market after 5 to 7 years following its introduction. Laser discs, he suggested, would become the initial medium. It was also stated that the 1125/60 line and field figures originated in Japan, but the 16:9 aspect ratio was developed by Dr. Kerns Powers of the David Sarnoff Research Laboratories in Princeton, NJ. Following his remarks generated at the IEEE 1988 Consumer Electronics Meeting in Chicago, Dr. Robert Hopkins, Executive Director of the Advanced Television Systems Committee, declared that everyone must be able to use a compatible ATV system—cable, satellites, and broadcasting—and that the FCC must offer priority for over-the-air spectrum, and should not reallocate UHF to land mobile until further study is made. "We need," he continued, "one clear, universal standard."

In a subsequent interview with William Hassinger, Assistant Chief of the Mass Media Bureau for engineering of the FCC, this gentleman expects today's television to evolve into HDTV in the form of a dual system with parallel transmission channels. And although the Zenith Spectrum Compatible HDTV system doesn't share a channel with conventional NTSC, a second channel doesn't bother the FCC. Engineer Hassinger recognizes the attractiveness of the proposed wide screen, notes 3-D effects are more pronounced, and that action scenes almost involve the viewer. He hopes the industry will pull off a miracle and converge on something worthwhile since this is a changing world that's wide open. Companies are very much alive, while cable worries about video tape recorders, broadcasters worry about cable and satellites, satellites worry about transponder leases and sales, and full specifications for the new service have yet to be completely written. Hopefully, the FCC's job will be completed in 1991, but could carry over into 1992.

In the meantime, let's investigate the FCC's initial thinking as a good clue to that which is yet to come. Guidelines, as you will observe, are relatively ambitious and fairly conservative.

MM DOCKET 87-268

Essentially, this "tentative" decision and notice of further inquiry is a review of Part 73-E of TV broadcast stations operational requirements and reevaluation of UHF channel and distance separation requirements, dated 9/1/88. Following this translation, we'll also include excerpts from a month earlier study of UHF receiver interference immunities, and any other HDTV-associated words of order or wisdom the Commission may have issued prior to print time. Therefore, much of the early reasoning and guidance will have been recorded for contemporary use or posterity, whichever seems appropriate. Throughout this chapter and the book, FCC abbreviations terminology ATV stands for *advanced television* and includes all the variously proposed disciplines now being considered for both TV audio and video.

The report finds that ATV broadcast techniques and existing broadcasters would benefit the public; that present spectrum allocations are currently sufficient; compatible NTSC service must continue; future systems requiring more than 6 MHz to deliver a non-compatible signal won't be authorized; but there will be no roadblocks in the way of independent ATV introduction in other services or the non-broadcast media. So CATV cable, video player/recorders, and satellite deliveries have relatively free hands—provided they wish to begin manufacture, programming, and distribution before the establishment of a North American Standard. This could involve considerable risk, but is just the ticket for R and D and demonstrations. In the meantime, the FCC will set no specific time for additional comments because of the uncertain timetable of developments.

During this interim period, the Commission has sought comments on recommendations and conclusions of the Advisory Committee, and further information on ATV terrestrial broadcast systems under design, including their abilities to operate under VHF/UHF interference limitations. Members also want to know how ATV standards should be established and whether to relax or repeal NTSC, "distribute supplemental spectrum," allotment and assignment procedures, and licensee negotiations with one another over service areas.

Meanwhile, new station applications and allotment requests have been FCC-frozen in 30 major cities where a spectrum shortage could exist if an extended ATV spectrum is approved over and above the 6 MHz now allocated to NTSC. In addition, the FCC will defer action on further spectrum sharing between UHF and land mobile until the next interim report is received and considered.

While all this is taking place, the commission will continue to cooperate with the Advanced Television Systems Committee (ATSC) as well as the Advanced Television Test Center (ATTC) once it is in operation, probably close to the end of 1989 or the beginning of 1990. The ATSC was formed in 1983 by the TV industry to coordinate and develop voluntary national technical standards for ATV systems.

The Advisory Committee has warned that long-term considerations and options may differ markedly from those in the short term and desires further evaluations as ATV progresses. At the moment it is indicated that a 9 MHz non-contiguous system compatible with NTSC is more spectrum efficient than dual 6 MHz channels for separate compatible and non-compatible signals. But long-term considerations might be just the reverse, if and when ATV establishes "significant" market penetration "taboos" can be relaxed or discarded, so that ATV systems may operate co-channel or adjacent channel without undue interference. If such conditions cannot be overcome, spectrum outside that already authorized for TV will have to be reviewed. And the ATSC also suggested development of inexpensive interface connections and/or receivers able to adapt to several standards.

Internationally, Great Britain and probably the rest of Europe will deliver HDTV by satellite rather than on the ground, in time for direct-to-home delivery by DBS and the MAC (multiplexed analog component) format. Special receivers for either satellite or cable will be required. Japan will also offer satellite-delivered HDTV, but also a terrestrial NTSC-compatible signal with characteristics somewhere between NTSC and broadband satellite transponders. MUSE HDTV, with a bandwidth of 9 MHz, is scheduled for sometime in 1990, and improved NTSC in 1989 having a bandwidth of 6 MHz. The Japanese claim their MUSE system could offer a gradual change (transition) from NTSC to HDTV. Our own David Sarnoff Laboratories in Princeton, NJ makes the same claim.

As for a world standard, ATSC and SMPTE (The Society of Motion Picture and Television Engineers) have already approved an interlaced standard of 1125 lines and 60 fields/second. While such an international standard is not within the FCC's "jurisdiction," the Commission, nevertheless says program signal format information of various system program availabilities are relevant and production to transmission sources do bear on final results in many instances. The FCC also wants comments on system compatibility with 35mm film standards, especially the 1125/60 and European 1125/50 fields/second proposals. During the interim, the FCC will continue to do its own investigations of compatible and incompatible NTSC systems already reported.

NTSC COMPATIBLE: A REVIEW

Hitachi says its system has an aspect ratio of 4:3 with progressive scanning using frame-store techniques and 3-D luma and chroma filters. This is claimed to reduce cross color/luminance artifacts and increase resolution of both brightness and color information.

Matsushita offers quadrature modulation of both detail and the 16:9 aspect ratio with the video carrier thereby adding an extra 1 MHz of additional bandwidth within NTSC for added detail. Matsushita, nonetheless, wants additional spectrum for further improved performance.

Faroudja Laboratories has a Super NTSC already in hardware form that largely removes cross color and undesirable artifacts.

Del Rey Group has developed subsampling to compress high definition pictures to 6 MHz within NTSC compatibility. Their tri-scan techniques divide a picture element into three subpixels that are then transmitted serially in successive frames. For stationary or fixed objects, the frame rate is 10/sec., but increases to 30/sec. for motion. Aspect ratios are 16:9, but 5:3 may be possible.

Nippon would divide the video signal between 4.2 and 5.6 MHz into blocks having assigned priorities. Quality images, they say, can then be efficiently transmitted with as little as 5-10% of an entire block.

Director Wm. Schreiber of MIT's Advanced Television Research Program uses a vertical portion of the NTSC frame to improve resolution and believes that subsampling could add enhanced color and digital audio.

David Sarnoff Research Laboratories with NBC backing is developing two formats: One would have subcarriers modulated in quadrature for enhanced resolution and sidepanels for a 5:3 aspect ratio. Called ACTV1, it can use either 1050 interlace or 525-line progressive scanning. ACTV II could deliver wideband video and digital sound but would need an extra broadcast channel in part or whole.

NHK is also working on MUSE-6 for single channel operation with an upgrade option requiring an additional 3 MHz on another channel.

VISTA, developed by New York Institute of Technology, does maintain NTSC compatibility but also needs an extra channel to improve both spatial and temporal resolution. NTSC signals are not changed, but frame rates for added information are somewhat lower to conserve bandwidth.

North American Philips, calling their system HD-NTSC, was originally a 12 MHz system but has been reduced to 8 MHz. Two channels are required: the first is standard NTSC and the second has sidepanels for an aspect ratio of 16:9, additional spatial and temporal resolution, "pan and scan" (wide angle selection for NTSC viewing's narrow angle) and probable digitization of the additional 3 MHz information. Narrow MUSE has also been developed by NHK, and is a system that is said to operate on a single channel but has only been computer-simulated thus far in late 1989.

NONCOMPATIBLE NTSC: A REVIEW

The following three systems are designed, according to the FCC, for satellite operations and are all noncompatible with NTSC.

Philips HDMAC-60 is a MAC system that uses time division multiplexing that is said to skillfully separate luminance and chroma to remove artifacts and deliver a DBS-type of signal directly to homes for consumer consumption, as well as businesses, too.

Scientific Atlanta has developed HDB-MAC for similar purposes and has stepped high definition of its business oriented B-MAC now in regular use throughout the U.S.

MUSE HDTV has band-limited luma and color difference information which is then sampled. Frame stores and motion detection in the receivers are said to compensate for any quality losses over regular MUSE.

OTHER CONSIDERATIONS

The FCC also wanted to know if over-the-air systems were of singular importance to U.S. communications and entertainment, and concluded they were. "Unlike many other countries," the FCC decided, "the United States has a strong and independent system of privately-owned and operated broadcast stations that transmit local and regional news, information and entertainment as well as national and international programs. Therefore, initiating an advanced television system within the existing framework of local broadcasting will uniquely benefit the public and may be necessary to preserve the benefits of the existing system."

Those two sentences may very well explain precisely why the Federal Communications Commission has chosen the broadcast medium as its HDTV yardstick. If there was no free programming and reception, many citizens would be denied the benefits of this new and potentially rich service that will deliver video and sound most people haven't yet imagined.

As former FCC Chairman Richard Wiley freely predicted before the Washington, D.C. IEEE Broadcast Group's annual luncheon, in the 1990s there will be large, flat-screened TV receivers with pictures on the wall of photographic quality, and HDTV will amount to a new frontier in broadcast and cablecasting. And if we don't take advantage of HDTV, the U.S. could become a second class electronic power. Conversely, attorney Wiley warned against a "suicidal" commitment to terrestrial broadcasting and said non-broadcast mediums such as cable should develop their own delivery means . . . and even coexistence with the telephone companies is necessary. Right now, he continued, we need laboratory testing and system evaluation as well as further spectrum analysis. "Standards," Mr. Wiley said, "are great, but let's not rush to judgment."

At a prior meeting of Electrical and Electronics Engineers, it was observed that the "flat panel people have a great deal to do before overcoming the cathode ray tube." To this we add a fervent amen, especially in the large flat panel category. However, this is far from impossible and advances are occurring almost daily, so you never know. Just don't hold your breath for the next 3 years.

THE SPECTRUM PROBLEM

ATV requirements, channel capabilities, system costs, industry preferences, technical advantages, supplemental allotments, possible interference, and optimum picture/sound quality are all FCC concerns that require both considerable attention and exceptional judgment. Final decisions here will largely determine transmission quality, receiver costs, economic impact, service beginning, and an American first or last system. These are not easy determinations since final decisions will affect commercial and consumer electronics industries of not only the U.S. but the world. Consequently, expanded spectrum usage, whether adjacent channel or separated, may be a long time in coming. Many tests and studies are necessary before final choices are made.

As of 1989-1990, most broadcasters say it's too early to make a useful decision on whether 6 MHz NTSC has sufficient bandwidth to compete generally with full HDTV on cable, VTR, or satellite systems. Neither the mood of the public nor enhanced picture quality will be known until systems are actually on the air. Therefore, some proponents are arguing that extra spectrum should be preserved/reserved to allow terrestrial HDTV to fully develop.

The Japanese are saying there is little probability of any 6 MHz system being developed that can compete in quality with their 9 MHz MUSE, but our own Zenith Corp. points out that color TV compatibility with black and white seemed an impossibility in the past and that in time, a 6 MHz bandwidth might very well offer "fully acceptable picture quality." The main difference between the two positions is now and then, and that's why the FCC is likely to proceed only with due deliberation unless someone develops a compatible system in a hurry. Zenith's, by the way, is non-compatible but uses only a single channel. An interim step between the two, of extended definition for terrestrial uses, seems reasonable before the 1990s are well underway.

HBO owner *Time* objects to bandwidths greater than 6 MHz for cable because of severe technical problems for cable distribution of broadcast signals. Broader bandwidths would necessitate new receiver converters and changes in headend equipment. *Time* also observes that if non-contiguous spectrum (non-adjacent) is authorized, cable could undergo significant ghosting and airplane flutter, as well as probable difficulties with harmonically-related carriers added to minimize interference.

The *Time* view is also supported by testimony from others, including Blonder Tongue, who says that in the real world of ghosts, noise sources, and atmospheric anomalies, non-contiguous channel additions from separated antennas could even reduce ATV picture quality below simulcasting (broadcast by ATV and TV at the same time) so that programming would not necessarily have to be identical on NTSC and ATV, and ATV could actually develop as a separate service.

The Planning Subcommittee of ATSC and its Spectrum Utilization and Alternatives Working Party came to two short-notice conclusions due to time limitations and fluid parameters of suggested ATV systems: 1) additional spectrum is more often available where channels are not continuous; and 2) there are performance/protection co-channel and adjacent channel combinations that could be suitable for all broadcasters (and allotments). To provide ATV additional channel would have to tolerate a 6dB D/U ratio, and that assignments would naturally be more difficult in urban rather than rural domains. Should such conditions not exist, the Working Party concluded all stations probably could not be accommodated or that existing service areas would have to be trimmed. A special report by Jansky and O'Connor to Working Party 3 declared the UHF taboos won't necessarily preclude ATV systems since receivers can be designed to reject any NTSC interference. Further, they said collocated transmitters offer better i-f beat protection and intermodulation interference than those having the usual 20-mile separation. The ARB augmented signal, they continued, would have none of the problems that led to taboos adoption in the first place. ATV augmentation, they argued, would be so low-powered there would be no UHF-NTSC interference. As to the 1-13 GHz spectrum, the Working Party finds a possible opportunity for ATV to operate on a shared basis in part of the range but that further considerations would have to await ATTC propagation tests.

TABOOS

In the late 1970s and early 1980s, the Federal Communications Commission became quite concerned over receiver interference problems in the assigned TV spectrum and issued a ruling designed to prevent co-channel and adjacent channel interference by directing certain exceptions to continuous channel assignments in specific locations. With the prospect of advanced television (ATV) and improvements in receiver tuner design, the FCC's engineering laboratory (presumably at Laurel, MD.) has made a further study and statistical analysis of 15 1983 receivers to "examine the impact ATV might have on the existing television receiver population."

UHF tuners are known to have problems rejecting interference in their currently assigned spectrum between 470 and 806 MHz. However, ATV "augmented" (meaning extra-channel) signals are expected to be specially designed to reduce interference to main transmitter signals and therefore have much less effect on extended or high definition receivers. The FCC doesn't have such sets available for testing just yet and characterizes the August 1988 efforts "as a useful first step in studying ATV interference." It identified "relative levels" of barely perceptible interference between tuned and taboo channels. The channels are identified as "U" undesirable taboo channels to "D" desirable tuned channels (D/U). The greater the ratio, the better the TV performance. The 15 receivers were those used by Laurel previously in earlier taboo research and are all electronically tuned, most being frequency synthesized. Observers there were said to be "expert viewers."

In all, 14 taboo channel spacings were at levels of -15 dBm (strong), -35 dBm (moderate) and -55 dBm (weak—a total of 42 evaluations for the 15 receivers. Excluded were taboo channels 2 through 5 since their intermodulation products are considered equivalent, and normality tests were computerized using Gaussian Probability distribution.

Table 2-1 summarizes the results, "N" being the number of the tuned channel initially, and $N + 7$ . . . would represent possible oscillator co-channel interference to another receiver tuned out $+7$ . . . this is so because the co-channel local oscillator usually operates 3.75 MHz above the lower portion of $N + 7$. An i-f beat, on the other hand is represented by $N \pm 8$ channels since two oscillator signals may combine to produce beats within the receiver's i-f amplifiers.

As for sound and picture interference, the FCC assigned $N \pm 14$ channels for sound and $N \pm 15$ channels for picture. The visual carrier of any picture is more vulnerable than the sound carrier, since sound carriers are traditionally lowered in amplitude to deliberately reduce possible interference between the two. The image channel band is located both above and below the local oscillator. Intermodulation and cross-modulation channels are identified as $N - 2$ and $N - 4$, respectively. Upper adjacent channel problems are numbered $N + 1$.

This table represents D/U signal ratio required to protect consumer-type receivers now operating at specific thresholds. Input signal intermodulation can provoke spurious signal within the tuned channel, whereas cross modulation

Table 2-1. Estimated Threshold of Undesired-to-Desired Signal Ratio Needed to Protect 90 and 50 Percent of the Receiver Population.

Undesired Signal	Desired Signal Strength		
	Weak (-55 dBm)	Moderate (-35 dBm)	Strong (-15 dBm)
Upper Adjacent Channel (N + 1)	(a)	*0dB/9dB	−6dB/−1dB
Lower Adjacent Channel (N − 1)	* −6dB/8dB	* −6dB/5dB	* −6dB/−1dB (b)
Intermodulation Channels (N − 2, N − 4)	* −16dB/21dB	10dB/14dB	−4dB/1dB
Intermodulation Channels (N + 2, N + 4)	*2dB/12dB	−2dB/6dB	−6dB/0dB
Cross Modulation Channel (N + 2)	17dB/25dB	8dB/17dB	−4dB/3dB (b)
Cross Modulation Channel (N − 2)	21dB/27dB	13dB/20dB	(c)
Cross Modulation Channel (N − 4)	30dB/36dB	(d)	(d)
Half - IF (N + 4)	(e)	−1dB/7dB	* −5dB/1dB (b)
IF Beat Channel (N + 7)	10dB/23dB	* −8dB/10dB (f)	* −14dB/0dB (b)
IF Beat Channel (N − 7)	6dB/22dB	* −2dB/13dB (f)	* −12dB/2dB (b)
IF Beat Channel (N + 8)	*5dB/21dB	*17dB/9dB	* −17dB/2dB (b)
IF Beat Channel (N − 8)	4dB/21dB	*5dB/13dB (f)	*10dB/2dB (b)
Sound Image Channel (N + 14)	−1dB/13dB	−2dB/8dB	−6dB/2dB
Picture Image Channel (N + 15)	−20dB/−7dB	−17dB/10dB	−26dB/−19dB

Notes:
*Data was conditionally normal

(FCC)

usually means the transfer of undesirable visual carrier information into the tuned carrier.

The study concluded that although power levels of ATV augmented signals would be 4-6 dB under that of their collocated primary transmitters, "a significant increase in interference to stations' primary service areas may be possible." Consequently, it was recommended that taboo channel interference protection be increased from the 50% receiver population figure established in 1952—this, they believe, would or should protect 90% of receivers from such interference. A −15 dBm reading, represents a worst case condition if the receiver is bothered at that level by interference. Study results indicate that adjacent channels, intermodulation channels, and sound image channels are best for collocated ATV augmentation channels.

Since this was an "eyeball" type of examination, one would fully expect that the Advanced Television Systems Committee and its ATTC laboratory will make careful "signal" examinations of these results.

OET SPECTRUM STUDIES

The Commission's Office of Engineering and Technology (OET) also undertook a pair of spectrum studies for possible ATV service. The first considered assigning extra ATV spectrum within existing U/V channels between 2 and 69, principally by eliminating UHF assignment taboo restrictions.

This study was conducted in two parts: No. 1 looked at adding 3 or 6 MHz of contiguous (adjacent) spectrum for systems planned for either 9 or 12 MHz; No. 2, doing the same for additional spectrum, which is not necessarily contiguous for either extra bandpass or channels. In this respect, separation between TV station assignments depends largely on locations and distances required to prevent interference. This usually correlates with the *desired/undesired* (D/U) ratios, transmission powers, and antenna heights and gains remaining constant. Minimum D/U ratios depend largely on perceived picture quality, antenna characteristics, and receiver interference rejection.

The availability study analyzed percentages of broadcast stations, applicants, and "permits" for possible added spectrum with minimum separations of 100 to 190 miles, although at 100 miles, ATV receivers would have to work with much lower D/U margins than regular NTSC sets. This would amount to D/U ATSC margins of 6-10 dB rather than the 28-45 dB of NTSC units. Adjacent channel distances of 60 miles separation were left unchanged. Of the 1760 station statistics examined, 706 were VHF and 1057 UHF. Interference protection for land mobile operations between channels 14 and 20, as well as existing Canadian and Mexican stations was considered, although additional spectrum for ATV for these non-U.S. stations was not. Low-powered and translator broadcasters were excluded also.

Co-channel and adjacent interference were prime portions of this study and assigned taboos *were not*. This is because the report considers that new ATV receivers can be designed and manufactured inexpensively to avoid taboos.

NTSC receivers versus taboos were not examined in the study. OET also says that the computer-simulated solutions do not necessarily represent the most efficient assignment of added spectrum, and that additional study is needed before specific spectrum can be assigned. The results of nationwide and major market simulations for ATV spectrum availability appear in Tables 2-2 and 2-3. As you will see, densely populated areas in near proximity would suffer the most, while widely separated broadcast locations would not be severely affected. Note distances and proposed additional spectrum. This could be an excellent key to many future allocations when and if the FCC acts.

Study No. 2 on possible receiver interference problems is little more than a repetition of that already outlined at the beginning of the "taboo problem." Consequently, these data and explanations will not be repeated.

TENTATIVE FCC SPECTRUM DECISIONS

While bypassing fixed satellite and DBS satellite bands, and generally avoiding all microwave regions, the FCC does believe that ATV systems can be designed for non-NTSC interference and not be susceptible to NTSC UHF taboos at distances "significantly less than those at which NTSC stations interfere with each other." Interested parties should plan for and design systems capable of working within the above limitations.

The FCC said, if ATV was limited to the present 6 MHz service, new receivers could pick up both NTSC and ATV transmissions and also "would

Table 2-2. Nationwide ATV Spectrum Availability.

Amount of Additional Spectrum (MHz)	Minimum Co-Channel Separation		Conditions
	100 miles	190 miles	
3	77	22	ATV spectrum contiguous to
6	63	17	station's existing channel.
3	94	50	ATV spectrum in same frequency
6	84	38	band, i.e., VHF with VHF, UHF with UHF, not necessarily contiguous
3	100	77	ATV spectrum anywhere in existing
6	98	61	VHF/UHF broadcast bands, not necessarily contiguous

Based upon the OET study, this table indicates the percentage of existing stations[102] that could be allocated additional spectrum for minimum co-channel separations of 190 and 100 miles. Also, only adjacent and co-channel restrictions were considered. UHF taboo restrictions were not considered.

Note: These results approximate an optimized solution through the use of heuristic techniques that were developed to attempt to accommodate the largest number of stations nationwide.

(FCC)

Table 2-3. ATV Spectrum Availability in Major Markets.

City	Present No. Stations	Number of Stations Accommodated			
		6 MHz		3 MHz	
		100 miles	190 miles	100 miles	190 miles
New York	12	5	0	6	1
Los Angeles	15	9	5	12	5
Chicago	13	9	2	10	2
Philadelphia	10	5	0	6	1
San Francisco	13	8	8	9	8
Boston	10	4	0	7	3
Detroit	7	3	0	4	1
Dallas/Ft. Worth	15	11	9	14	9
Washington, D.C.	10	6	0	8	3
Houston	11	9	8	11	8
Nationwide	1760	1480	673	1655	881
		84%	38%	94%	50%

Note: Results are illustrative only. Priority could be accorded the top ten markets without significantly changing the percentage of stations accommodated nationwide, but the availability of spectrum in nearby markets would be affected. For example, if additional stations in the Washington, D.C. market are accommodated, there would be less spectrum available in the neighboring Baltimore and Annapolis markets.

(FCC)

permit future assignment of additional full-service television stations, would not displace LPTV and TV translator stations that operate on a secondary basis, and might provide an opportunity for additional non-broadcast use of the UHF band.''

This would permit economies in production that could lower receiver prices, and more ATV programming would increase demand and accelerate broadcasters and cable systems to install ATV hardware. But the FCC rationalized that it isn't clear if broadcasters can remain competitive eventually with a 6 MHz system.

An additional 3 MHz could allow a station to offer NTSC on its primary channel and ATV augmented information on the second channel for ATV-equipped receivers. But broadcasters would have to upgrade transmitters, cable operators likewise, in addition to having the problem of adding capacity or eliminating some on-going services. Further, added spectrum could reduce many TV service areas, with the greatest reductions occurring in large cities. And supplemental ATV operation may not be practical where there are large frequency differences between prime and supplementary channels, especially among broadcasters in the lower VHF channels.

An additional 6 MHz channel could have two possible scenarios: The first would allow expanded signals or split transmissions, with NTSC on the second. The FCC initially seems to favor the second since after ''a transition period'' if

ATV wasn't selling, broadcasters could abandon this second channel and resume NTSC in full service areas. The FCC has asked for additional comments on this suggested ruling. Fewer broadcast stations could accommodate the dual system, however, especially in the major cities.

Effects are of considerable concern not only on TV stations, CATV, consumers, and spectrum efficiency but in the simulcast type system. And although additional spectrum may not be possible for existing broadcast stations in such cities as New York, Los Angeles, Philadelphia, Detroit, and the District of Columbia, limited channel reassignments could be considered for a few stations that would permit additional distribution of the ATV spectrum. On this matter, OET is now doing distribution of the ATV spectrum.

The FCC plans to conclude its technical analysis "quickly," develop a "variety of channel assignment plans," and present them for public comment as soon as possible. "At this juncture," the commission continued, "we see little benefit in deferring spectrum decisions until we reach a decision on technical standards issues." Early band decisions can identify hardware constraints that could narrow the number of systems offered and help designers develop ATV formats useful in actual broadcasting. If so, the FCC suggests such action might allow relaxation of our freeze on TV allotments and assignments.

Similarly, the commission refused to either amend or repeal the UHF freeze or any further sharing of spectrum with the private land mobile radio service. Once a variety of channel allotment plans have been adopted, however, members could repeal or relax such restrictions if in the public interest.

Satellite links are also under consideration if ATV needs more than 6 MHz. Considered are possible C-band in the Fixed Satellite Service, the Ku band, and even the 12.2 to 12.7 DBS downlink which may or may not ever fly. Even higher bands are also available. The Commission observes that a 25 MHz channel could be sufficient for a 9 MHz ATV transmission. Ku-band transponders are "typically" 54 MHz and could be split for dual video transmissions. Members would like to know the effects on such distributed signals to both broadcasters and cable systems. We would also have to consult our Mexican and Canadian neighbors if this new technology generates more interference than the present 6 MHz system. At the U.S.-Canadian boundary, the proximity of densely populated areas could be a problem. Canada is already participating in ATSC meetings, and Mexico is expected to join the deliberations eventually.

THE QUESTION OF STANDARDS

Among the many other problems requiring detailed consideration are those of standards. The NTSC standard was codified by the FCC in 1953 and is used as an example. The FCC must now decide if NTSC should be relaxed or repealed, how to establish standards for ATV, and if there should be compatibility among ATV transmissions and any other ATV sources such as cable and/or VCRs. Current testings and evaluations now underway should prove a determining factor as further results are analyzed and tabulated. During this period,

NTSC will not be relaxed although temporary and experimental transmissions will take place from time to time—all within existing Commission Rules. Waivers are expected occasionally from the FCC granting the broadcast of ATV signals if there's no interruption of NTSC or interference with other NTSC transmissions.

The FCC is taking under advisement all inquiries to modify NTSC regulations that might improve signal quality. Many of the HDTV proponents among broadcasters ask the FCC for a single mandatory ATV transmission standard, while others with newly-designed transmission systems either oppose such action or aren't quite sure. On-going testing should resolve most divergent views in the near future. The FCC will take a strong stand in the final decision, "with the advice and involvement of all sectors of the industry."

The FCC believes that a standard will stimulate investments in ATV technology as well as further public acceptance since receivers will be compatible with transmitted signals. Any such standard, the FCC feels, should not restrict future development of ATV with subsequent improvements.

There are some alternatives to a single, mandatory standard, and the FCC is also examining these:

1. Protect some standards by prohibiting any interference to that standard, but not requiring its use. This might allow development of additional systems with better features and at less cost.
2. Adopt some standard for "allocation and assignment . . . only" and allow future flexibility to ATV modifications.
3. Permit a standard to be optional after a "sunset" period. A limited standard effective only for a specified period. The FCC also would like to look at costs in developing an open architecture receiver capable of decoding at least two or more standards. But the Commission admits that arguments for this course of action are not well developed and more technical and cost information is needed.

The FCC would like comments on an "optimum time" for standard adoption. Meanwhile, resources and expertise of the industry will be crucial to successful development of such a standard. And so industry cooperation from such as the American National Standards Institute (ANSI), the Advanced Television Systems Committee (ATSC), and the Electronics Industry Association (EIA), are encouraged.

RECEIVER COMPATIBILITY

According to the Commission, almost all industry participants agreed that ATV and NTSC compatibility for receivers is desirable, but one comment continued to advocate retention of the simulcasting provision with inexpensive converters. Others would opt for NTSC on one channel and 3 MHz or 6 MHz supplemental carriers on another. And the third option seems to be NTSC on one channel and a totally incompatible signal on another.

The FCC believes that maintaining NTSC is very important and there should be "no dislocation" of existing TV broadcast service. Commissioners believe ATV signals should either be compatible with NTSC transmissions or that ATV stations simulcast NTSC with ATV. Should these views continue to prevail, NTSC will remain on the air for the 140 million TV sets now operating, with ATV eventually being available over broadcast stations for those who purchase new receivers capable of receiving both signals.

Now the FCC would like to know if broadcasters would continue NTSC without special regulations; would NTSC quality levels be required if ATV is also mandated; and if ATV-to-NTSC converters could be built reasonably, should ATV still be received on regular receivers? Obviously these are extremely important considerations for the entire manufacturing industry due to cost, technical improvements, and engineering times. For it will necessarily be sometime between final Federal decisions and the actual introduction of the receivers.

Industry respondents generally agree that for cable, satellite services, VCRs, and alternative video media, NTSC compatibility is desirable rather than attempting adapters, and a common ATV standard would aid compatibility among the several systems since costly conversions could easily degrade image quality.

It may be likely that ATV transmission methods could vary among the various services due to "different tradeoffs between bandwidths and picture quality." Baseband (detected) video/audio signals may be preferable to some common denominator TV system since cable CATV converters and satellite receivers can be designed to supply broadband information directly to TV monitors and avoid any "transcoding" of ATV signals among different formats as well as typical losses resulting from remodulation. For those not informed, representative video cassette recorder/players and satellite receivers already deliver audio/video baseband outputs, including stereo sound. Satellite receivers can also deliver a detected composite signal 8 MHz in bandwidth, and that's approximately what the better TV/monitor combinations can generally process. The only thing they can't do is change aspect ratios from 4:3, 5:3 or 16:9, but that could possibly be overcome, too, in time.

For feeder and distribution systems, the Advisory Committee's Planning Subcommittee has already recommended a multistandard capability of RGB (red, blue, and green) inputs plus digital sound. This would also be advantageous for Teletext and MAC-B. The Electronic Industries Association has adopted an EIA IS-15 Multiport Standard to accommodate NTSC and peripheral devices, and this same standard will also allow luminance/chrominance (Y/C), color difference RGB-Y and red, blue, and green RGB inputs. In RGB, brightness is derived from the three primary colors and is not transmitted separately. In extended Y/C definition formats, luminance and color are received separately. There are already decoders that transpose Y/C to RGB. Therefore, the foregoing proposals represent only current technology and nothing especially new. EIA has recently modified its proposal and now endorses RGB-Y inputs because of the Y/C extended definition VCRs and camcorders currently available. Color difference information can be matrixed for Y/C.

If others change their views also, one of the three inputs may not be recommended after all. But in the final analysis, one of the three will become the preferred receiver medium, especially since Teletext and B-MAC are transmitted RGB, but B-MAC originates as R-Y, B-Y, and G-Y in color difference serial modes at the transmitter. It is also true that most television receiver systems deliberately process chroma and luminance separately and then combine them again shortly before they enter the three guns of the cathode ray tube. In the better receivers, a comb filter separates the two after the video detector. France's SECAM television system does broadcast color difference information, but video is frequency modulated and audio amplitude modulated—just the opposite of our vestigial (one sideband suppressed) dual sideband AM modulation video system. How the various manufacturers will respond to any "universal" input is not known and probably won't be until a general system standard has been recommended or adopted by Federal authority.

The FCC observes that while a number of receiver makers in the CATV, VCR, and satellite group support several varieties of intermediate coding, they all are compatible with 525 line/59.94 Hz fields via either baseband or NTSC RF. And this may prove an incentive for manufacturers to design one or more inputs that will become universal. And the same may be said for ATV compatibility among satellite, VCR, and CATV media without government intervention.

The FCC wants more inputs on the subject, both pro and con. For instance, is compatibility among various video products in the public interest; should the Commission mandate or recommend some standard; is an "open architecture receiver" a viable alternative to some voluntary standard; name the costs and benefits; and does the commission have legal authority over compatibility standards for non-spectrum electronics such as VCRs?

There's more than one problem and possibly a number of fairly reasonable answers for each. We thought you should have some idea of what a regulatory agency must contend with as it sorts through chaff and facts before rendering either an opinion or regulation. Some determinations are not especially easy nor broadly popular, but they must be made for the good of all, and not simply for certain special interests whose argumentative voices are often the loudest.

ON-GOING CONSIDERATIONS

The FCC fully expects that nationwide ATV distribution will be both complex and expensive. All licensees probably will not proceed at the same rate, and some spectrum may lay idle as the demand for ATV develops. The Commission therefore, is considering any unused supplemental spectrum interim use for non-ATV purposes. This would be limited to the ATV transitional period and ancillary uses. Nor would such "ancillary" uses by broadcasters or others be permitted to interfere with either ATV or NTSC transmitters.

The Commission, therefore, will encourage beneficial technical change and permit broadcasters to assimilate ATV technology, with the objective of improving existing broadcast service. And the FCC wants to assign additional spectrum to "existing licensees and applicants" because of their already heavy investments and expertise. But members also recognize some legal and policy reasons to allow spectrum applications from others who have a special 6 MHz TV service.

As of now, any additional spectrum needed would be assigned among existing VHF/UHF channels, not necessarily adjacent, depending on prior assignments. In very crowded areas, some VHF stations may have no supplemental spectrum with which to work.

During the approach period, the FCC could specify ATV criteria and applicants could apply. But a deluge of applications to overwhelm the Commission might cause more harm than good. Therefore, a second method might be to allocate all spectrum at once, matching 3 and 6 MHz supplementals with existing NTSC grants, and a new Table of Allotments issued in a single ruling. But preferred choices of some broadcasters might not be satisfied, and some stations might not be accommodated at all.

According to the Commission, a third approach could combine certain elements of the first and second considerations in a two-step method by giving as many stations as possible supplemental spectrum, then undertake the second procedure possibly by private agreements, lotteries, or hearings. This would resolve many anticipated disagreements and resolve any excess grants in certain areas. Licensees would probably be given a fixed amount of time to "use it or lose it." And some broadcasters may decide not to use ATV now or in the foreseeable future, and decline any supplemental assignment. Reassignment of some NTSC channels to provide better ATV flexibility might be necessary. Some station licensees may want to exchange service area characteristics to solve an audience or technical problem.

Obviously there are a multitude of considerations outstanding that the electronics industry and the FCC will have to resolve before final action is taken. But the FCC still plans to first use the existing TV spectrum exclusively, retain NTSC at least during the transition period, develop allotment plans, require no delay in ATV implementation for other services and the non-broadcast media, but remain sensitive to compatibility of associated equipment.

In the meantime, the Commission wants more information on ATV systems now under design for broadcast ATV, especially technical characteristics and bandwidths. Comments are sought on standards and there should be a ruling on equipment compatibilities between broadcast and non-broadcast gear.

I thought it worthwhile to include a list of those filing Initial and Reply Comments to offer some idea of the extent to which HDTV affects the U.S. If you know video electronics, you'll recognize the names of many who are absolute leaders in the broadcast and TV industry. The listings will also clear up any unexplained abbreviations appearing in text we may have overlooked.

As you can see, this is one gigantic undertaking that affects so very much of the industry and even the entertainment world. Everyone, we're sure, hopes only for success.

FILINGS TO THE FCC

A-Vision, Inc. (A-Vision)
American Family Broadcast Group, Inc. (American Family)
Association of Independent Televisions Stations, Inc. (INTS)
Association of Maximum Service Telecasters (MST)
Black Television Workshop of Los Angeles (BTW)
Blonder Tongue Laboratories (Blonder Tongue)
Bonneville International Corp. (BI)
Broadcasting Technology Association, Japan (BTA)
Bundy, Jr., Walt W. (Bundy)
Capital Cities/ABC, Inc. (Cap. Cities/ABC)
CBS, Inc. (CBS)
Center for Advanced Television Studies
Chronicle Broadcasting Co., (Chronicle)
Corporation for Public Broadcasting, National Association of Public Television
    Stations, and Public Broadcasting Service (Public Broadcasters)
Cosmopolitan Broadcasting Corporation (Cosmopolitan)
Cosmos Broadcasting Corp. and M&C Communications, Inc. (Cosmos)
Cox Enterprises, Inc. (Cox)
David Sarnoff Research Center, Inc. (Sarnoff)
Del Rey Group (Del Rey)
Digideck, Inc. (Digideck)
Dolby Laboratories (Dolby)
Electronic Industries Association, Consumer Electronics Group (EIA-CEG)
Electronic Industries Association, Satellite Communications Section (EIA-SCS)
Faroudja Laboratories (Faroudja)
Fisher Broadcasting, Inc. (Fisher)
General Electric Consumer Electronics Business (GE)
General Instrument Corporation (GI)
George N. Gillett, Jr. (Gillett)
Great American Broadcasting Co., McGraw-Hill Broadcasting Co., Inc. and
    The New York Times Company (Times Broadcasting)
Hearst Corporation (Hearst)
Hitachi, Ltd., Central Research Laboratory (Hitachi)
Hughes Communications Galaxy, Inc. (Hughes)
Japan Broadcasting Corporation (NHK)
Japan Satellite Broadcasting (Japan Satellite)

King Broadcasting Corporation (NHK)
Land Mobile Communications Council, Drafting Committee (LMCC)
Matsushita Electric Corporation of America (Matsushita)
Meredith Corporation (Meredith)
Metrovision, Inc., Newchannels Corporation and Sammons Communications, Inc. (Metrovision)
Motion Picture Association of America, Inc. (MPAA)
National Association of Broadcasters (NAB)
National Black Media Coalition and the NAACP (NBMC)
National Broadcasting Company, Inc. (NBC)
National Cable Television Association, Inc. (NCTA)
National Captioning Institute, Inc. (NCI)
National Public Radio (NPR)
National Telecommunications and Information Administration (NTIA)
Neuman, W. Russell (Neuman)
New York Institute of Technology (NYIT)
Nippon Television Network Corporation, Engineering & Technical Operations (Nippon)
North American Philips Corporation (NA Philips)
Outlet Broadcasting, Inc. and Atlin Communications, Inc. (Outlet)
Post-Newsweek Stations, Inc. (Post-Newsweek)
Pulitzer Broadcasting Company (Pulitzer)
Radio New Jersey (Radio New Jersey)
Radio Telecom and Technology (RTT)
Radio-Television News Directors Association (RTNDA)
Rogers Cablesystems of America, Inc. (Rogers)
Satellite Broadcasting and Communications Association of America (SBCA)
Schreiber, William F. (Schreiber)
Scientific Atlanta (Scientific Atlanta)
Scripps Howard Broadcasting Company (Scripps Howard)
Time, Inc. (Time)
Times Mirror Broadcasting (Times Mirror)
Toshiba America, Inc. (Toshiba)
Tribune Broadcasting Company (Tribune)
United States Advanced Television Systems Committee (ATSC)
Viacom International, Inc. (Viacom)
Walt W. Bundy, Jr. (Bundy)
Zenith Electronics Corporation (Zenith)

Reply Comments to the FCC

Association of Maximum Service Telecasters (MST)
Association of Maximum Service Telecasters, National Association of Broadcasters and the National Cable Television Association (petitioners)
Association of American Railroads (AAR)

Broadcasting Technology Association, Japan (BTA)
CBS, Inc. (CBS)
Corporation for Public Broadcasting (Public Broadcasters)
Cox Enterprises (Cox)
David Sarnoff Research Center, Inc. (Sarnoff)
Del Rey Group (Del Rey)
Hitachi, Ltd., Central Research Laboratory (Hitachi)
Japan Broadcasting Corporation (NHK)
Land Mobile Communications Council, Drafting Committee (LMCC)
National Association of Broadcasters (NAB)
National Assn. of Public Television Stations and Public Broadcasting Service
    (NAPTS)
National Broadcasting Company, Inc. (NBC)
National Cable Television Association, Inc. (NCTA)
North American Philips Corporation (NA Philips)
Radio Telecom and Technology, Inc. (RTT)
Satellite Broadcasting and Communications Association (SBCA)
Sony Corporation (Sony)
Station Representatives Association, Inc. (SRA)
Televisa, S.A. (Televisa)
Wireless Cable Association, Inc. (Wireless Cable)
Thomson Consumer Electronics, Inc. (Thomson)
Time, Inc. (Time)
VisionAire, Inc. (VisionAire)

3
The Major HDTV
Systems Proposed

AT LAST COUNT THERE WERE MORE THAN 15 "PROPONENTS" OFFERING extended definition (EDTV) and/or high definition (HDTV) systems for television industry and FCC approval, one or more that could be approved by the Federal Communications Commission. This chapter contains a detailed outline or outright analysis of each of these major offerings, depending on available source material, some extensive and others sketchy. Most of the respondents are depending on either Dolby or Digideck to supply digital audio. These two sound specialists will be included in chapter 4 with special emphasis on encoding techniques and significant parameters.

The objective of each of these proponents is similar: to produce the best possible image within a single or dual-channel passband and deliver maximum bandwidth efficiency—the ratio of picture quality to RF bandwidth. Methods vary, such as brute force FM for satellite transmissions, information companding (compressing and expanding), or matching optics and electronics to properties of the human eye. All are represented among the various schemes, although terrestrial systems normally have some semblance of NTSC compatibility as required by the Federal Communications Commission. The dual channel arrangements usually augment an NTSC-receivable channel with 3-6 MHz of expanded bandwidth or special effects such as progressive scanning and 16:9 (width-to-height) CRT aspect ratio display compared with NTSC's standard 4:3. There is also at least one "open architecture" receiver suggested to receive and reproduce video/audio for all systems. The FCC has already expressed interest in such a unit. If one was available, the Commissioners might conceivably avoid picking a single HDTV source and avoid possible lawsuits from the losers. But if no one system enjoys the Commission's blessing, then HDTV could very well become embroiled in an AM stereo-type fracas that would set the entire "video revolution" back an easy 10 years. For everyone concerned—the gainers and losers—let's hope that a similar fiasco does not reoccur again. Considering the FCC's ratio of lawyers to engineers in recent years, that currently deregulating body should be able to handle legal anything.

MIT-RC AND MIT-CC

MIT-RC and MIT-CC are the systems proposed by Wm. F. Schreiber and Andrew B. Lippman, Massachusetts Institute of Technology's Research Technology and Media Laboratory. They are described together and in sequence, since similar principles mostly apply; one is simply an extension of the other, and they are referred to as a "family" by their designers. System philosophy, according to Dr. Schreiber, should be a concept that includes the entire Tx/Rx chain from camera to display since advanced television systems to be introduced will continue operating into the next century using sophisticated receivers. Schreiber and Lippman therefore assume an advanced receiver will not only have frame store and a flicker-free display but also substantial computer (computational) power. A simple receiver for HDTV "implies" parallel red, blue, and green channels at a total bandwidth between 30 and 50 MHz scanning in sync with the camera and doing little signal processing. But being limited to one NTSC channel and one additional "augmentation" channel of either 3 or 6 MHz won't exactly pass 30-50 MHz, therefore other methods are obviously necessary to produce a high definition picture within these limitations.

The MIT approach would eliminate a separate carrier, the retrace intervals, and vestigial sideband transmissions. Instead the proponent would quadrature-modulate two 3-MHz signals on a single carrier at band center, with "various components" being time multiplexed, leaving some one-twelfth of the time for a combination of audio and data, which might be either digital or best quality FM.

Camera, channel, and display could operate with separate scanning standards and applicable frame stores. A high line-rate progressively scanned camera and display allow very good vertical resolution without interline flicker, although additional effort may be needed to overcome excessive flicker on interlaced displays. At the same time, Schreiber and Lippman criticize 60 fps as a poor choice with a flicker-free picture. They suggest the motion picture 24 fps rate and much-improved film interlace, along with receiver motion compensation to permit very smooth images even down to 12 fps (on film). Higher spatial and lower temporal resolution is also suggested, as illustrated in Fig. 3-1.

Evidence indicates a diamond-shaped cross-sectional frequency response in three dimensions will save considerable bandwidth. There is some object-blurring at slow, steady speeds, but not a great deal more than the product of camera integration. Another advantage is that sound quality is improved when added to film at the 24 fps rate for mixed frame-rate systems. Then, a smart receiver could select the best frame rate for each scene depending on motion. Thus a limited rate set based on multiples of 12 fps should produce high spatial

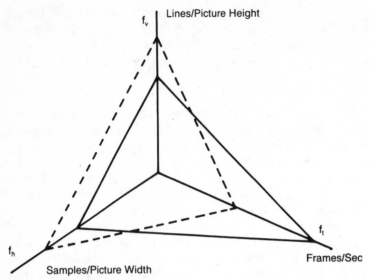

Fig. 3-1. Vertically-oriented rectangular frequency response whose volume equals channel capacity (courtesy MIT R and D Laboratory).

resolution in slowly moving scenes and higher temporal response with rapid motion. As for color, MIT says that psychophysical studies suggest chroma visual bandwidth is about half that of luminance, NTSC color seems to be overdesigned, and that color can be successfully transmitted at 12 fps. Taken collectively, the number of picture samples/frame would amount to a factor of five or six over NTSC in stationary scenes.

Signal/Noise Improvements

In all audio/video communications systems the one supremely important parameter is signal-to-noise (S/N) that's always directly related to quality pictures or sound. If 54 dB broadcast studio and 56 dB satellite S/N characteristics continue to be standard for baseband video, then maximum effort has to be directed toward program origination, transmission, and reception since any one component can adversely affect the whole.

Channel noise suppression, for instance, would be the use of isotropic narrowband RGB lows and luminance highs, since highs require reduced S/N compared to lows. Further, extra bandwidth would be available for enhancement in addition to different frame rates for spatial-frequency transmissions and receptions.

Dr. Schreiber confirms that the weakest link in the transmitter-receiver chain appears in TV over-the-air or cable signal transmissions, and further advocates the "mixed highs" system with its adaptive modulation. This, he declares, raises signal levels in the highly visible blank areas and reduces noise at the receiver. In his system, only RGB information fills the lows, while others are high, and some noise may even be shifted from busy blank portions by companding. Adaptive information, he states, requires little channel capacity and must be transmitted for isotropic effect.

A Smart Receiver

A smart receiver adaptable to the MITV-CC cable and MITV-RC 6 MHz systems, comparable to Japan's MUSE in cost and complexity, but with open architecture and better programmability, would receive an improved image in an NTSC-compatible system and offer a ''bridge'' to ultimate HDTV. The EDTV receiver, however, would offer improved spatial resolution on the new system but old receivers would see a 16:9 aspect image with top and bottom ''bars'' due to height-width tradeoffs.

A block diagram of this receiver appears in Fig. 3-2. Both input and image reproducing portions are fixed, but the computerlike processing section can be

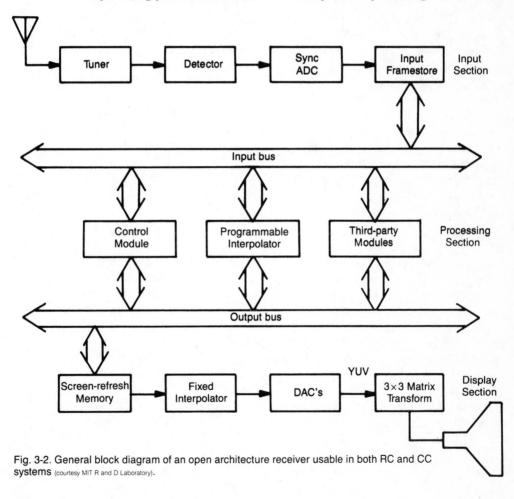

Fig. 3-2. General block diagram of an open architecture receiver usable in both RC and CC systems (courtesy MIT R and D Laboratory).

programmed by minor digital commands within the main signal. The top input portion consists of tuner and RF amplifiers, the detector, sync processor, analog-to-digital converter (DAC), and first framestore. Here, a digital representation of the original frame as transmitted, but in nondisplay form, must pass through the center portion with an input and output bus in addition to the control and third-party modules and the programmable interpolator. The buses are interfaces between input electronics and also accommodate any third-party modules needed for upgrading or possible interfacing with digital inputs such as fiberoptics. The processing portion typically receives input frame store intelligence, rearranges and prepares it for display memory, all under supervision of the control module by either manual or data commands from outside or originating transmit sources. The processor is said to be made up of low-level digital hardware that should become inexpensive in quantity and cheap to program.

MIT states that such a programmable receiver could easily adapt decoding parameters to different TV transmissions, including NTSC, and would allow further additions such as motion-compensated temporal interpolation, echo cancelling, and noise reduction. It might also invite additional plug-ins for software or hardware modules and other specialized inputs. Outputs for such a receiver include a screen-refresh memory, fixed interpolator, digital-to-analog (DAC) converters, an RGB output format, and a 3-×-3 matrix transformation prior to the cathode ray tube.

The MIT system does not overlook diamond-shaped spatial frequency schemes of competing systems with vertically-oriented rectangular frequency responses. And if there's proof positive that this is a truly useful arrangement, separable filters may be installed and scans rotated 45°—an undertaking that is impossible in analog systems we know today.

NYIT VISTA SYSTEM

Already operating in hardware, the New York Institute of Technology in Dania, Florida, claims its system is fully NTSC-compatible, and uses a single wideband or two regular NTSC channels. It may be set up as a regular 6 MHz channel plus another of equal bandwidth; two 6 MHz channels time-shared by two stations; or 6 MHz plus 3 MHz in the augmentation channel. Based on the assumption that optimum eye resolution of any display amounts to some 22 cycles/degree of visual angle and that contrast sensitivity is approximately 30% of peak sensitivity, designer Dr. Wm. Glenn observes that seven screen heights for NTSC and three and one half for 1125-line HDTV are common viewing distances—a 2:1 reduction. Consequently, Dr. Glenn considers that regular NTSC of 60 fields per second and 2:1 interlace is sufficient for low resolution motion but a bi-channel system with reduced frame rate for added luminance detail in an (extra) augmentation channel would produce an 8 MHz display from its original 30 MHz camera or taped input.

Vertical resolution is usually limited by the number of active scan lines of the picture, while *horizontal resolution* relates directly to the video signal's bandwidth; thus the accepted equality that 1 MHz represents 80 lines of X-axis resolution. There are also two other types of resolution called *spatial* and *temporal*; the first applying to picture detail *without* motion, and the second to motion accuracy. Spatial resolution can be determined by a 0.7 Kell factor and the static or dynamic interlace factor vertically, while horizontally calculations involve active line duration times the compressed video bandwidth times an inverse of the aspect ratio. Temporal resolution is considerably more subjective since it directly involves motion, but is most dependent on the number of fields needed for complete receiver, horizontal and vertical resolution. To improve all picture-perceived elements, the 4:3 NTSC horizontal viewing ratio is expanded to 16:9 and possibly converting 525, 2:1 interlace vertical scans of 262.5 each into so-called progressive scan which contains all 525 lines in every field. This doubles the number of scan lines but also doubles the bandwidth.

With these explanations now complete, let's continue with Dr. Glenn's VISTA. He would temporally filter detail information to dispose of those frequencies near or above half the frame rate by either integrating the camera tube image or by using frame stores. Consequentially, frame rates as low as 7.5 frames per second (fps) can supply sufficient motion detail in the augmentation channel. As for another aspect ratio, say 5:3, this system could expand the 525-line video information into the horizontal line blanking interval by 4 μsec, shortening vertical by 10% but expanding horizontal by 10%—simply a tradeoff.

In the 1125-line system, signal detail from the progressively-scanned camera reads into frame store, and NTSC images are scan-converted to the 1125-line format by line interpolation and added to the progressive format readout as shown in the illustration. A block diagram of the VISTA transmitter (Fig. 3-3) follows, along with a similar non-detailed group of rectangles for the VISTA receiver (Fig. 3-4).

For the transmitter you see both 1125-line progressive scan and 525-line NTSC luma-chroma inputs. The upper input is first digitally filtered with mid and high frequency video detail processing. The high Y is routed directly to a summing amplifier and there joined by 1125 progressive scan conversion as well as frame index and burst. The single output then proceeds to the augmentation transmitter. The mid Y detail is routed to a 525 interlaced scan processor and then into a second amplifier to be joined there by A/D-converted Y luminance from the 525 NTSC interlaced signal augmented by a temporal enhancer. Color difference R-Y and B-Y information is next delayed, temporally filtered and applied to the encoder for NTSC baseband transmission in addition to the Y luminance.

The receiver diagram shows both baseband and augmentation "select code" tuners, whose signals are again A/D-converted for frame store and then converted to progressive scan for the digital filter (top) and frame comb filtered

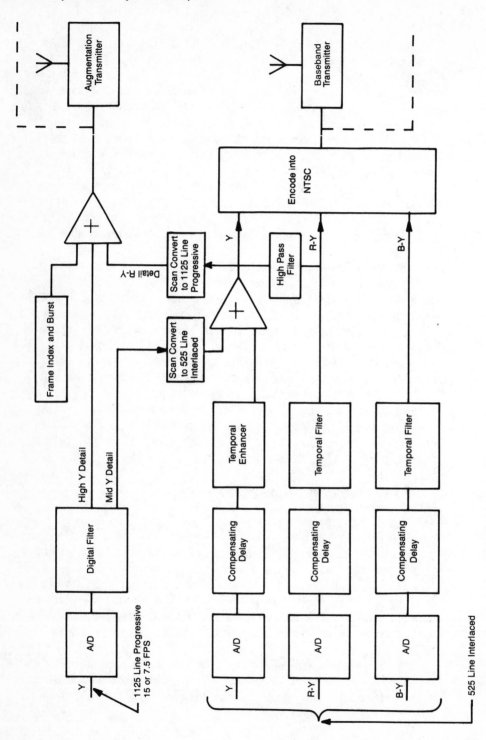

Fig. 3-3. Block diagram of the duly digitized VISTA transmitter.

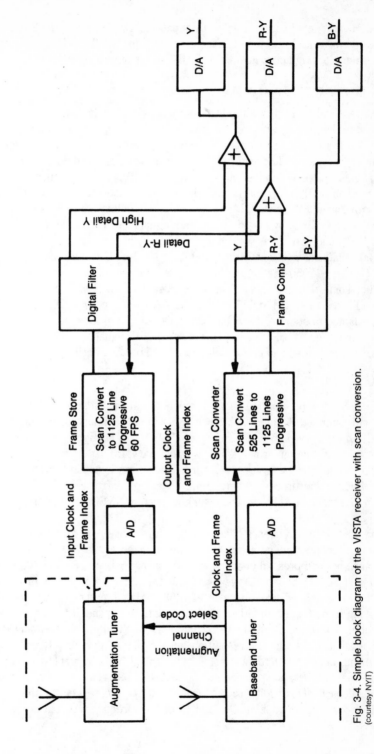

Fig. 3-4. Simple block diagram of the VISTA receiver with scan conversion.
(courtesy NYIT)

below. High Y detail and R-Y detail then pass to two separate amplifiers that receive the original Y and R-Y, in addition to B-Y for the three D/A converters and resultant analog luminance and R-Y, B-Y color difference outputs.

SARNOFF'S ACTV-I and ACTV-II

ACTV-I and ACTV-II are products of the David Sarnoff Research Center in Princeton, New Jersey and were demonstrated publicly for more than a year before hardware system testing in computer simulation. Now fully integrated, Advanced Compatible Television as ACTV-I can be considered improved NTSC for standard 4:3 aspect ratios, but will expand to 16:9 width and height dimensions with an ACTV receiver in addition to better horizontal and vertical resolution. Introduced as a single 6 MHz/channel compatible system, ACTV-I can be expanded to ACTV-II by encoding augmenting signals in a second channel that offers high definition TV as well as digital audio enhancements.

Both units are specifically designed for a 1050-line, 2:1 interlaced system, even though a 1050-line 59.94 fields/second, progressively scanned model would be somewhat more ideal. This is because a 1050/2:1 line and aspect ratio can readily be converted to 525/1:1 (progressively scanned) for ACTV-I, and is easily adapted to standard NTSC at 525/2:1 (interlaced) for ordinary transmissions. Such rates would preserve all regular NTSC signal and sync properties, but permit certain additions that will extend audio/video performance.

ACTV-I

In ACTV-I, four separate signal components are included for wider aspect ratio and extended definition, three of which reside within "subchannels" of NTSC. In ordinary 63.5 μsec line scan, the central portion of the picture in widescreen-origination format converts to ordinary 4:3 aspect ratio, increasing the horizontal bandwidth since widescreen was time-expanded to fill all but 2 μsec of active line time. And since time expansion narrows bandwidth, widescreen luminance and chrominance will now occupy NTSC parameters (Fig. 3-5).

Two horizontal frequency bands are generated, one μsec at the end of each active line accepts time-compressed low frequencies, while 700 kHz to 5 MHz luminance not compressed in overscan and chroma between 83 and 600 kHz is prefiltered in 3-D horizontal, vertical, and temporal prefilters, with chroma quadrature-modulated on luminance at 3.58 MHz. Together, sidepanel pairs occupying some 6 μsec on either end are now time-expanded to fill the center line panel. These are the first two components, with another two to follow.

For additional horizontal resolution beyond the 5 MHz already contained, frequencies between 5-6 MHz are shifted 0.1 to 1.1 MHz by amplitude modulation with a 4.9 MHz suppressed carrier and low-pass filtered to remove the upper sideband. This additional luminance is then compressed for the center panel, recovering an additional 1.1 MHz of high frequency detail across the

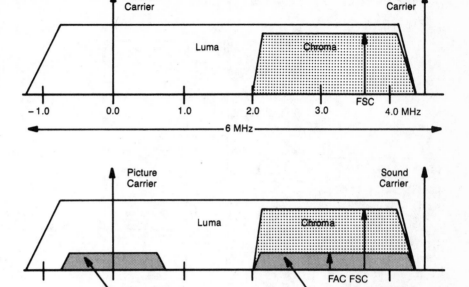

Fig. 3-5. Standard NTSC RF spectrum (top) and ACTV RF spectrum (bottom) showing location of extra modulated components (courtesy David Sarnoff Princeton Research Laboratories).

entire image. In the ACTV-I receiver, the 1050/2:1 signal downconverts to 525/1:1 (progressively scanned) format, permitting a "helper" signal to add additional vertical-temporal luminance for better vertical motion detail, and is one of low average energy. It's horizontally-filtered to "about" 750 kHz. Other kinds of helper signals are also under investigation (Fig. 3-6).

Receiver Signal Recovery

The two prime considerations in reception/decoding are system compatibility and intelligence recovery. Any additional components hidden in the main signal must remain so on regular NTSC receivers, with all components recoverable in the ATC receivers without crosstalk (Fig. 3-7).

Compatibility requires sidepanel low frequencies (high energy) hidden in horizontal overscan, with low energy information amplitude-compressed and quadrature modulated on a new 3.1 MHz subcarrier interleaved at an odd multiple of half the line-scanning rate and phase inversion on alternate fields. The 750 kHz helper is spatially correlated in the center picture panel and removed on receivers with synchronous detectors and "perceptually hidden" on sets with envelope (usually diode) detectors.

Components 1-3 are recovered by a process identified as *interframe averaging*—a linear, time-varying digital filter arrangement that separates modulated

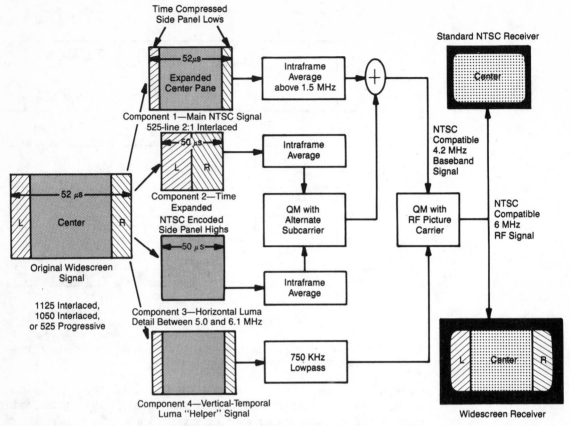

Fig. 3-6. ACTV—A single channel NTSC-compatible widescreen EDTV system (courtesy David Sarnoff Princeton Research Laboratories).

and baseband signals free of vertical-temporal crosstalk, even when there is motion. Horizontal crosstalk disappears with guardbands between horizontal and pre-post filters.

There is also an ACTV-E somewhat-reduced performance option using a 525-interlaced widescreen format rather than progressive scanning or 1050-line HDTV resolution. It will, however, retain a wide screen via the sidepanels and the added subcarrier. The ACTV-I enhanced resolution is retained and home receivers will display a progressively-scanned picture from components 1, 2, and 3. In effect, this is a low-cost option of ACTV-I, less any expanded vertical resolution in moving scenes.

When the four encoded portions have been recovered and converted to luminance Y and I-Q components, widescreen decoding occurs. The center picture panel must be compressed to $4/5$ of an active line, sidepanels need to be re-expanded, and high frequency sidepanel detail also needs compression to its original time slot. When all relevant signals have been placed in widescreen format, extended horizontal luminance information shifts to its original frequency

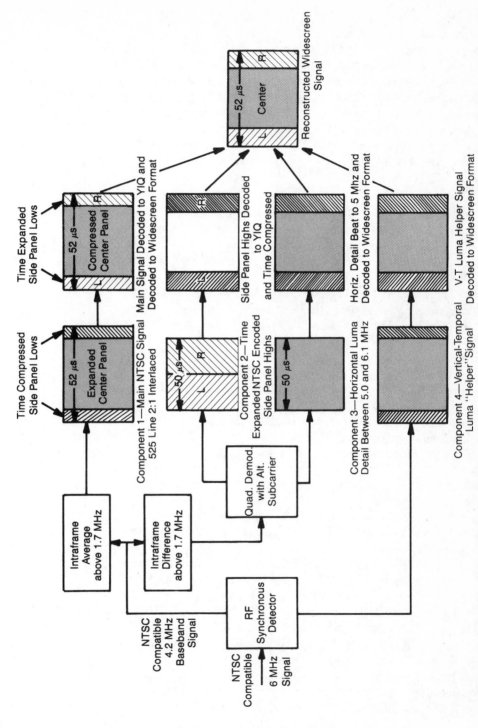

Fig. 3-7. Pictorial block diagram of ACTV decoder (courtesy David Sarnoff Princeton Research Laboratories).

range by 4.9 MHz carrier amplitude modulation and is then added to the main luma intelligence for a passband of 6 MHz. This interlaced signal converts to progressive scan with temporal interpolation and helper assist, while chroma converts to progressive scan, but with no temporal interpolation. Lastly, Y and I-Q intelligence is re-converted to analog and matrixed for RGB output.

Sync information for these receivers transmits in one of the subchannels discussed. For digital processing, a pseudo-random signal sequence prompts the receiver to do a correlation and offers timing accuracy in nanoseconds to identify the first pixel in each field. Ghost reduction is also required, as well as the basic identifier for either ACTV-I or II transmissions.

ACTV-II

ACTV-II is the real HDTV 2-channel system but accepts complete ACTV-I transmission characteristics and expands them. The ACTV-II encoder contains an ACTV-I encoder, and this is the prime 6 MHz signal going to ACTV-I and NTSC receivers on the compatible channel. But ACTV-II also encloses an ACTV-I *decoder* to recover widescreen Y-IQ and the 525-line progressive scan format. Y-IQ in 12 MHz and 1.2 MHz (each) bandwidths, respectively, are then subtracted from ''the original'' HDTV to produce augmentation signals of Y at 20 MHz and I-Q of 10 MHz each. In this operation, progressive scan information is subtracted from interlaced information, with augmentation signals now containing all spatial-temporal details not transmitted by ACTV-I in addition to necessary artifact corrections in decoded ACTV-I (Fig. 3-8).

The 20 MHz Y information subsequently splits into three bands: 0-6 MHz, 6-11 MHz, and 12-18 MHz, with the remainder not transmitted. Middle and upper bands are ''shifted'' to dc by 6- and 12-MHz carriers, respectively, while the low band is designated Y_L. The two upper bands are then low-pass filtered to $1/2$ Nyquist vertically within each field. Resulting information is now line-multiplexed for a 6 MHz output termed Y_H. In moving picture portions only Y_L transmits, while in still scenes, Y_L and Y_H are frame multiplexed. Switching between the two results from a binary motion signal in the helper signal or from luminance frame differences in ACTV-I. Compressed luminance 6 MHz information ''spans'' 754 pixels (picture elements) at $8\times$ the color subcarrier frequency and proceeds to the luma/chroma/data multiplexer.

The 10 MHz I and 10 MHz Q intelligence is horizontally low-pass filtered to 2.4 MHz, chroma is vertically low-pass filtered to $1/2$ Nyquist within a field and are line-multiplexed into one 2.4 MHz block. Compressed by $5\times$ horizontally, it now emerges as 12 MHz, separated into 0-6 MHz and 6-12 MHz bands, the lower band identified as C_L and the high band—shifted to dc by a 6 MHz carrier with upper sideband rejected by a 6 MHz low-pass filter—is known as C_H. In moving scenes, only C_L transmits, while in still portions, C_L and C_H are frame-multiplexed. Switching signals are the same for luminance, and 6 MHz chroma occupies 151 pixels at $8\times$ the color subcarrier, which is also routed to the luma/chroma/data multiplexer.

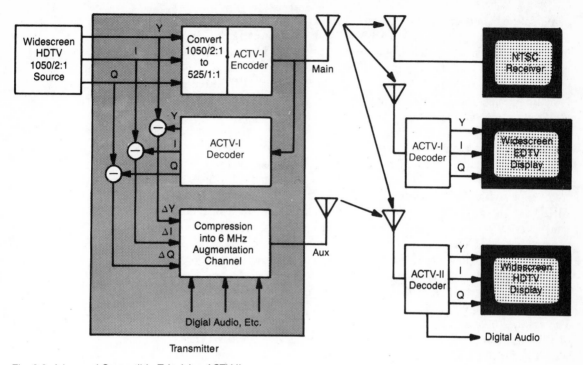

Fig. 3-8. Advanced Compatible Television ACTV II (courtesy David Sarnoff Princeton Research Laboratories).

Augmentation, established at 1050 lines interlaced and 59.94 fields/second, is sampled at $8 \times$ the 3.579545 MHz subcarrier and produces 910 samples/line. Of the 1050 lines, 960 carry video, leaving the rest for digital sound and data. Five pixels in video are reserved for horizontal sync. As for RF modulation, the 6 MHz augmentation information at baseband can be divided into even and odd-line data with time expansions by $2 \times$, resulting in two 3 MHz transmissions. They are then analog-converted and quadrature-modulated on the center of the RF channel.

ACTV-II Receiver

The ACTV-II receiver is tuned to the appropriate ACTV-I subchannel, identifies the first pixel in each field, and the augmented channel is equalized to match the main channel. Odd and even lines become quadrature demodulated and are compressed/multiplexed into one frame.

Both Y and IQ information must now be expanded so the receiver can produce its advertised passband. Initially they are demultiplexed and then processed separately.

In luminance motion the Y_L is directly added to the decoded ACTV-I luma information. But where there is no motion, Y_L and Y_H are frame demultiplexed and frame repeated in forward time for Y_L, and backwards for Y_H. Y_H is line

49

demultiplexed into signals that are vertically interpolated by line averaging and moved into original frequencies by 6 MHz and 12 MHz carriers. The three bands then combine to produce an augmentation signal added to the ACTV-I Y intelligence. This permits resolution of 18 MHz of horizontal resolution at low vertical frequencies, and 960 lines of vertical resolution under the same conditions. In moving scenes, 6 MHz of horizontal resolution appears with maximum vertical and temporal resolution.

I and Q expansions have similar treatment but, of course, at much lower frequencies. Once again, only C_L reaches the receiver in moving scenes while being added directly to ACTV-I decoded chroma, following time expansion plus I/Q demultiplexing. In the still portions, C_L and C_H are frame demultiplexed and frame repeated, time forward for C_L, and backward for C_H. C_H beats down to 6 MHz and adds to C_L. Following time expansion and I/Q demultiplexing, both are added to ACTV-I decoded information. In consequence, 2.4 MHz horizontally can be resolved at low vertical frequencies with 240 lines of vertical resolution, and 1.2 MHz of horizontal resolution at 240 lines of vertical resolution along with full temporal resolution.

There's more to this explanation, but the foregoing should offer enough background and foreground for a reasonable understanding of what the Sarnoff Laboratories have so far accomplished. Whether 1050 lines versus 1125 lines and 59.94 Hz or 60 Hz remains something to be decided later. Nevertheless, EDTV and then HDTV appear, at the moment, to be the high definition path to the future.

3XNTSC BY ZENITH

3XNTSC is a radically different HDTV system proposed by Zenith, and is of considerable interest to both industry and the Federal Communications Commission. Developed primarily for cable TV systems, it is also transcodable to FM for satellite transmissions, and offers improved threshold characteristics for better signal-to-noise ratios. Zenith also claims that with reduced average and peak power requirements, 6 MHz HDTV transmission over cable is readily achieved without rebuilding CATV facilities or decreasing the number of available channels.

Based on a scanning rate of three times the 15,735 Hz horizontal frequency of NTSC, the system offers a 1575-line, 2:1 interlaced picture that can be converted to other formats and "can carry 1125/60 line/field information in a 16:9 wide aspect ratio window." Signal inception requires a pickup device of 37.8 MHz to deliver a 29-30 MHz picture (Fig. 3-9) consisting of 1449 active lines/ frame at a horizontal scan rate of 47,203 Hz. Zenith compares these figures to the 1125/60 "standard" HDTV system which permits 725 lines of picture height and 1288 lines of width—assuming a Kell factor of 0.7, that relates apparent resolution lines to actual scanning lines. For transcoding to NTSC, "simple" interpolation filters would convert the 787.5 progressive line format, which might also be resolved as either NTSC or Y/C. As for magnetic tape or satellite formats that use frequency modulation, a time-multiplexed format such as MAC instead of the usual I/Q carriers would serve. Motion conditions are said to be comparable to NTSC except that any vertical edges and interlace lines would be at triple frequency, and much less discernible. Zenith would also develop analog and digital tape recorders they say would produce 35% better sound than those now under development for the 1125/60 line/Hz system. Zenith states that required transmitter power for their system would only require 0.2% compared to today's average transmitter, resulting in considerable hardware reduction, antenna gain, and/or antenna height.

HOW IT WORKS

Operating within the existing format of a standard NTSC 6-MHz channel, Zenith has found that high frequency components in video require less than 1% of the total energy, while almost all power needs reside in low frequency video and synchronization plus average or dc values.

By removing these power-consuming portions from the video portion of the signal, high frequency information can be transmitted with very little carrier power, better S/N ratios than NTSC, and almost no interference to other channels in proximity—actually less than 1% of the total Xmtr energy. Consequently, all video information above 200 kHz continues to be transmitted regular analog, while video below this frequency will be digitized, as will sync and dc components with low rates. Together with two channels of digitized CD-quality audio, these power-hungry elements are then digitized and assigned to the vertical blanking interval of some 22 lines for media Tx and Rx. Essentially, Zenith says their engineers are taking 30 MHz of composite video and audio, and compressing them into a 6 MHz channel with no effect on surrounding channels. Accordingly, average transmitter power needs are reduced by more than 90% over same-service area NTSC coverage. We might add that error correction circuits eliminate any secondary images or other distortions. And since a great deal of the low frequency information in NTSC is redundant (repeated), average values and sync intelligence do not have to be sent continuously, and require sampling at a very low rate, such as 11.988 fields/sec (fps) for video.

For video encoding (Fig. 3-10), the HDTV source is transmitted as suppressed carrier modulation on a pair of quadrature carriers at channel's center without subcarriers but with related NTSC timing so that spectral components can be properly interleaved. The HDTV signal is converted into 480 analog components per 1/59.94 seconds, with each component occupying 63.5 μsec at a nominal bandwidth of 2.675 MHz. These are now paired and transmitted sequentially as 240 pairs in suppressed carrier AM modulation on two quadrature carriers in the center of the RF channel, along with data and sync signals in the vertical blanking interval, consuming 22-23 scan lines on successive frames. Encoding permits a 59.94 fps rate for low frequency components, and an 11.988 fps rate for high frequency components. Red, blue, and green (RGB)

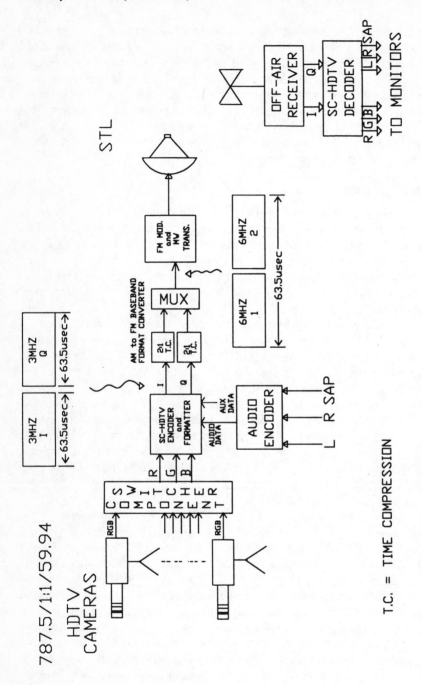

TERRESTRIAL STUDIO BLOCK DIAGRAM OF ZENITH SC-HDTV SYSTEM

T.C. = TIME COMPRESSION

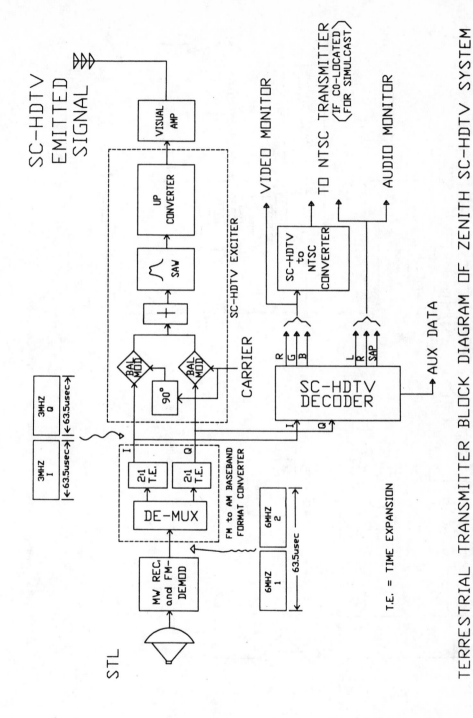

TERRESTRIAL TRANSMITTER BLOCK DIAGRAM OF ZENITH SC-HDTV SYSTEM

Fig. 3-9. System illustration of Zenith's 30 MHz full spectrum RF compression system used either alone or simulcast on a companion channel with **NTSC** (courtesy Zenith Electronics Corp.).

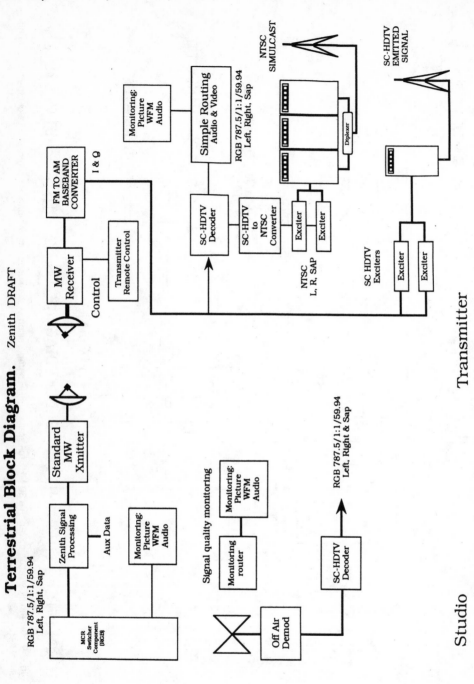

Terrestrial Block Diagram. Zenith DRAFT

RGB 787.5/1:1/59.94
Left, Right, Sap

Studio

Transmitter

Fig. 3-10. Block diagram of the high definition video encoding section of 3XNTSC (courtesy Zenith Electronics Corp.).

54

information is now separated into luminance (Y) and color difference (R-Y and B-Y) signals that Zenith describes separately.

Luminance Encoding

Here, luminance is divided into four components representing different frequencies in the two-dimensional (2D) spectrum. One set represents frequencies between 0 and 9.6 MHz and a full 720-line vertical resolution. Another, the middle portion between 9.63 to 19.27 MHz and 480 lines of vertical resolution, designated LL – LD, MH, HH, and C1, C2 for color, and the third grouping takes care of those frequencies between 19.27 and 28.9 MHz, displaying 240 lines of vertical resolution. Lowest vertical resolution line frequencies between 0 and 96 are encoded at 59.94 fps, with others time-multiplexed and encoded at 11.988 fps. Frequency bands are vertically filtered and resampled in the various resolutions for transmission. The majority of motion intelligence being dispatched among the 0-9.63 MHz horizontal frequencies at 1/59.94 seconds.

For bandwidth conservation, only active video is transmitted with no intervals between lines, and the higher speed-generated low frequency lines are time expanded 3.6:1, reducing their bandwidths to 2.675 MHz and expanding duration to 63.56 μsec. Temporal pre-emphasis is then applied to reduce interference, and low frequencies below 200 kHz are digitized and processed as data during the vertical blanking interval with no dc component and a much lower power requirement.

Color Encoding, Sound and Sync

Color difference intelligence is bandlimited to 9.63 MHz, vertically filtered and resampled, then divided into two components and paired into 240 resolution lines each and multiplexed down to 1 in 5, with active parts time-expanded by 3.6:1—much like luminance. Therefore, 96 color difference components are transmitted in pairs each 1/59.94 seconds, occupying times equal to 48 lines of NTSC information. Color difference resolution, however, amounts to only 1/3 of luminance.

All this produces 480 lines at 63.56 μsec duration and 2.675 MHz luma/chroma components requiring transmission every 1/59.94 seconds in pairs, leaving 22-23 lines available for data. This consists of two channels of compressed digital audio, low frequency video information, sync signals, captioning and error protection, transmitted in 16-state quadrature amplitude modulation (16 QUAM) on video-modulated carriers. I and Q data clock rates are set at 5.34965 MHz and flat to the Nyquist slope. Two lines carry sync signals along with error protection, permitting other remaining vertical blanking interval lines to carry 20,910 bits in 1/59.94 seconds. This also allows 8,334 bits to accommodate two channels of either Digideck or Dolby digital audio.

Video Decoding

In the receiver, the two pairs of 240 analog video and data information are detected, resulting in a 787.5-line progressively scanned high definition picture. A tuner converts the incoming 6 MHz RF into an IF passband where it is gain-controlled and a frequency and phase-locked loop (FPLL), establishes carrier regeneration for two quadrature carriers used to demodulate the I and Q intelligence. This is then digitized by a pair of A/D converters and proper ratios between analog and digital D/A information maintained by AGE and a variable gain amplifier. Analog information on the I and Q signals are now detected, digitized, and time-multiplexed into the four original luminance and two color difference components, effectively producing 720 active lines of progressively-scanned luma/color signals at RGB.

Luminance Decoding

As is customary, detection and repositioning of the various portions of the picture, sound, and sync, occurs in substantially the reverse to transmission. Digitized low frequencies are first reinserted and then line averaged, while other information is only line averaged; and by using an identical low-pass filter to that in the transmitter, precise restoration occurs without artifacts. Each of the frequencies from LL to HH now undergoes temporal de-emphasis with the various memory lines added, as well as some frequency shifting, in addition to interpolation filters for the 9.6-19.3 and 19.3-28.9 MHz groupings (Fig. 3-11).

Following temporal de-emphasis, the various frequency groups still have incorrect vertical display lines so they must be interpolated back to 720 active lines/frame with interpolation filters. Components transmitted at 11.988 fps are then combined and motion-compensated before recombination with the active lines. Receiver motion compensation being controlled by frame differences in 96/59.94 LL, also combine with motion compensation contained in the vertical blanking interval. As for color decoding, the same techniques are used for the 720-active line/frame display.

Zenith added considerably to this paraphrased description, but you should have enough here to reasonably understand the Rx/Tx system and its RF carrier, high frequency clock, and field rate sync details, as well as the necessity for AGE (automatic gain equalization) which automatically adjusts for zero carrier offsets and gain changes.

NORTH AMERICAN PHILIPS HDS-NA

North American Philips, owned and operated by N.V. Philips of the Netherlands—a global electronics manufacturer and marketer with 360,000 employees and 402 factories in 60 countries—is also offering a pair of HDTV systems, one for terrestrial broadcasts and a second for satellite transmissions. Under the composite designation of HDS-NA (High Definition System for North America), but then separated into satellite feeder signals called HDMAC-60 and HDNTSC for terrestrial delivery, these systems ''will produce 16:9 aspect ratios, pictures

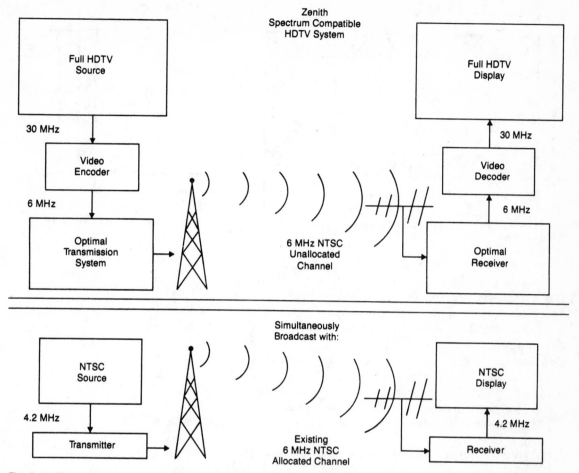

Fig. 3-11. The decoding portion of Zenith's 3XNTSC broadband HDTV system (courtesy Zenith Electronics Corp.).

free from motion artifacts, flicker, and line structure of more than 400,000 picture elements, in addition to four channels of CD-quality sound,'' according to Dr. Mark Rochkind, President of Philips Laboratories.

The system is said to be NTSC-compatible, will pan and scan, and much is already in hardware as of 1989-1990. Further CATV tests and satellite transmissions are scheduled for later in the year, probably after June. The HDNTSC cable/terrestrial format requires an additional 3 MHz augmentation signal in a second channel for maximum quality picture and digital sound, with the main channel remaining standard NTSC. Augmented, HDS-NA will deliver $1^1/_2$ times the horizontal and vertical resolution of NTSC in addition to wide aspect ratio and CD-quality sound, yet cause virtually no interference since augmented signal energy is considerably below standard transmissions.

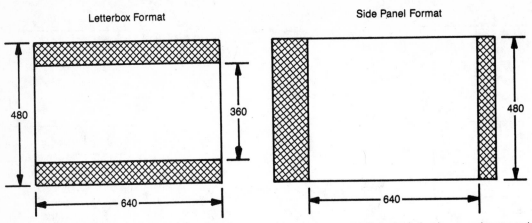

Fig. 3-12. Letterbox and sidepanel formats used in picture size and augmentation resolution in several proposed HDTV systems (courtesy North American Philips).

Reportedly, the system produces 563 lines if the signal source is 1125 lines; 525 lines for 1050 lines interlaced; and 525 lines for 525 lines progressive scan. This compares to 262 lines with the 525-line interlace input of standard NTSC. Added information and sound are included in the two "expanded" sidepanels illustrated in Fig. 3-12. Shown are both the HDMAC-60 and HDNTSC relationships and the extended resolution package. HDMAC, of course, is strictly a satellite method of transmission and is not used terrestrially since satcasting conveys frequency modulation and not analog. It is very important, however, since testing over Hughes Communications Galaxy satellite(s) has been planned (and probably executed) for the middle or fall of 1989.

More revealing than the block diagram is Philips system characteristics as set forth in Table 3-1. This indicates that HDMAC-60 can be used as a satellite feeder, DBS, and analog or digital fiberoptics. And HDNTSC is available for both CATV and terrestrial broadcasts. Bandwidth for both systems is listed at 16.8 MHz, and before transmission at 9.5 and 4.2 MHz, respectively, the latter being for main channel NTSC. The RF bandwidth for HDMAC-60, however, occupies 24 MHz. For modulation, the satellite means is FM, while the main channel for HDNTSC is the usual NTSC AM vestigial sideband, with augmentation channel either digital or analog at something probably less than 6 MHz.

Also note that with 525 vertical scanning lines as source, the HDTV resultant will become 525 progressive and 1050 interlaced at the display, with a field rate of 59.94 Hz. Vertical and horizontal picture resolutions for both systems are 480 and 495 lines, respectively, whether stationary or moving.

Production Formats—Terrestrial

The actual production formats currently proposed in its ATV presentation before Working Party 1 of the Systems Subcommittee for HDS-NA terrestrial/CATV, required 1050/59.94/2:1 interlace and/or 525/59.94/1:1 (progressive

Table 3-1. Philips HDS-NA System Characteristics.

SPECIFICATION	HDMAC-60	HDNTSC
APPLICATIONS	SATELLITE FEEDER DBS ANALOG OR DIGITAL FIBEROPTICS	TERRESTRIAL BROADCAST CATV
BASE-BANDWIDTH (MHz) *SOURCE SIGNAL *BEFORE TRANSMISSION	16.8 9.5	16.8 4.2 (MAIN CHANNEL) 3 (AUGMENTATION CHANNEL)
RF BANDWIDTH (MHz)	24	TWO CHANNELS: (6+3) MAIN CHANNEL: 6 AUG. CHAN: 56 (TARGET: 3 MHz IN THE N±1 ADJACENT BAND)
MODULATION	SATELLITE: FM	MAIN CHANNEL: AM-VSB AUG. CHAN: DIGITAL OR ANALOG
TYPE OF SCAN	PROGRESSIVE	PROGRESSIVE
# OF SCAN LINES/FRAME *SOURCE *TRANSMISSION *DISPLAY	525 OR 1050 525 525 PROGRESSIVE 1050 INTERLACE	525 OR 1050 525 525 PROGRESSIVE 1050 INTERLACE
FIELD RATE = FRAME RATE (Hz) *DESIRED SOURCE SIGN. *TRANSMISSION SIGNAL	59.94 59.94	59.94 59.94
PICTURE RESOLUTION BOTH FOR STATIONARY & MOVING (TV LINES/PICTURE HEIGHT) *VERTICAL *HORIZONTAL	480 495	480 495
SOUND *WITHOUT THE USE OF VBI *WITH THE USE OF VBI	2 DIGITAL MORE DIG. CHANNELS	2 DIGITAL MORE DIG. CHANNELS
COMPATIBILITY WITH NTSC	EASE OF (HD) NTSC TRANSCODING PAN & SCAN CONTROL DATA DELIVERY TO HDNTSC	FULL COMPATIBILITY PAN & SCAN

scan), said by Philips to produce less motion artifacts at 525 lines, while 1050 lines would work better with 30/24 frames/second film processing. The 60 Hz rate was not mentioned.

Similarly, the letterbox format with its decreased picture size top and bottom was discarded for sidepanel aspect ratio augmentation due to decreased picture size and diminished vertical and horizontal resolution for standard TV. In the sidepanel version, the picture size for NTSC is undisturbed, picture resolution unchanged, and pan/scan electronics allows HDTV distribution to remain software compatible for an NTSC audience.

Illustrating that the system can be expanded linearly with additional power versus bandwidth and no spectrum folding, both 525/1:1 and 1050/2:1 spectral distributions are illustrated in Fig. 3-13. Baseband frequencies of 16.8 MHz are the same, but the interlace and line subtraction contributions permit vertical resolutions of 680 lines for 1050 versus only 480 for 525.

As suggested previously, HDS-NA terrestrial is a two-channel system, with the original 6 MHz NTSC channel remaining intact and an augmentation channel at either VHF or UHF added for bandwidth and greater expansion. The only question outstanding on the additional channel is whether it should occupy 4 MHz and be frequency and time multiplexed; 3 MHz and frequency and time multiplexed; or 3 MHz and simply frequency multiplexed. Apparently it's significant to actually have three likely choices for positive HDTV expansion. However, since the prime principals are all embodied in the newest 3 MHz proposal, this will be selected for our basic explanation, once generic encoder and block diagrams have been illustrated and briefly discussed.

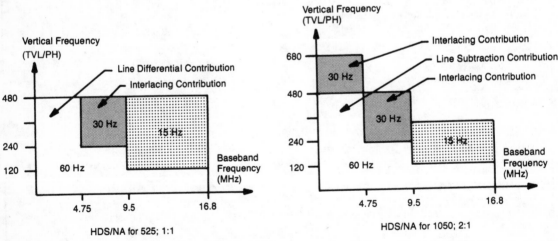

HDS/NA Luminance Spectral Distributions

Fig. 3-13. Linear expansion both vertically and horizontally (courtesy North American Philips).

Encoder/Receiver Blocks

Shown in Fig. 3-14 is the encoder channel with its RGB input matrix and Y-IQ translation. The line difference generator derives from luminance and thence on to the expanders, while Y continues into the bandstop filter and high/low frequency splitter, with highs traveling directly to the expander, and low luminance stopping at the center panel time split, where it is joined by IQ from the vertical prefilter. Also acting on center panel electronics is the pan and scan feature and this, too, controls both panels and line difference signals as well as a higher frequency luminance input via the augmentation channel formatter that accepts digital audio processing too. All of this enters the vertical interval blanking period (VITS) which outputs both augmented and main channel information.

Returning to the center panel above, its outputs are controlled color and low frequency luminance required by various expanders. Expander outputs are now shown as a line difference signal, high frequency luminance, along with composite panels and a composite center. How all this occurs will be explained immediately following the receiver illustration.

In the receiver block diagram (Fig. 3-15), you will note two pairs of RF inputs, one for the main channel and the other for second channel augmentation. Both inputs are demodulated and the results proceed directly to their

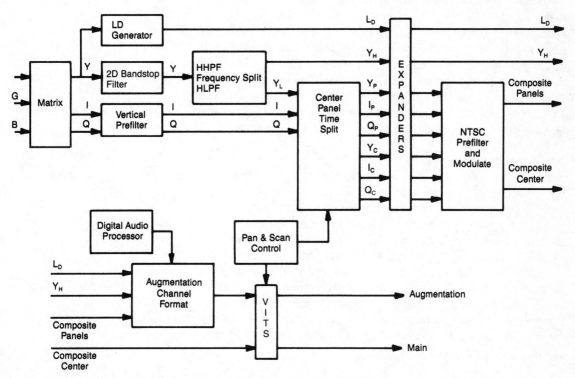

Fig. 3-14. RGB and expanded definition outputs plus pan and scan (courtesy North American Philips).

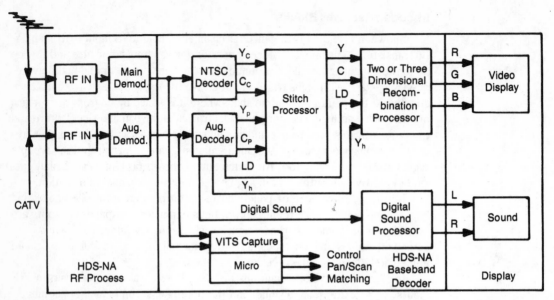

Fig. 3-15. Dual channel inputs and dual video/sound decoding in a generic receiver (courtesy North American Philips).

respective decoders, and then to the side panel stitch processor in both luminance and color. But there are also main and augmentation demodulator samplings for the vertical interval (VITS) capture circuits, along with MICRO electronics for control, pan/scan and raster matching. The augmentation decoder also supplies additional feeds for the digital sound processor as well as line difference signals and high frequency luminance to the two/three dimensional recombination processor that now receives separated luma and chroma intelligence from the stitch processor. Once these elements are again integrated, the output becomes RGB video and L/R analog sound. With audio quadrature phase shift keying (QPSK), 2 bits/Hz can produce 440 kilobits/second in augmentation FDM with video. In vertical interval augmentation, 2 bits/Hz supports 500 kilobits/sec—an interesting comparison between frequency domain multiplexing (FDM) and treatment in the vertical blanking interval.

Now that you have a broad outline of the system, some of the finer points of operational theory need review. The system selected will be the 3 MHz augmented channel version since this seems to be the safest and possibly the most comprehensive of prior proposals. If an additional 6 MHz RF channel is required to transmit a system's maximum video/audio response, the 1 MHz difference between 3 and 4 MHz is possibly inconsequential, unless some split-channel arrangement becomes necessary to conserve tight spectrum. Such may not be entirely feasible, however, due to relatively low augmentation emissions and the possibility of intrachannel crosstalk.

System Description

This two-channel proposal at the transmitter and recombination into a wideband display at the receiver is called bandsplitting. It is accomplished by transmitting low and high frequency components in two signals, the lower frequencies requiring most of the power, and high frequencies a much smaller portion. Here, the augmentation (extra) channel transports high frequency video, side panel expansion, and often digital audio in 2/4, L/R or 2(L/R) outputs, while the NTSC segment carries the rest. Sounds simple, but there are outstanding complexities.

In the 4 MHz augmented system using frequency multiplexing, NTSC and an additional channel do not have to be adjacent. Components in the second channel are frequency multiplexed and then combined via quadrature in pairs with the exception of Yh high frequency information. All must be companded (here expanded) to accommodate 127.11 μsec, double our usual 63.5 NTSC line scan time. Certain correlated signals with original bandwidths of 3.3 MHz are compressed to 0.68 MHz and then remodulated in quadrature on a subcarrier. Other signals aid pan and scan, and Yh will occupy 1.2 MHz after suitable expansion.

In the 6 + 3 (MHz) somewhat less complex expansion, quadrature modulation can accommodate some of the augmented information on the video carrier of channel number 1, but only with the addition of an inverse Nyquist filter at the transmitter to eventually resolve pure NTSC at the receiver for channel 1. Correlation is also needed to prevent crosstalk between any two quadrature-modulated signals. Here, several carriers are required and this is not always desirable. Therefore, another 6 + 3 version has been designed by Philips, a mixture of time division multiplexing and frequency division multiplexing. Apparently, this unit overcomes several prior objections, and reputedly will gain hardware status for planned general demonstrations. Time multiplexing, however, does restrict expansion, but central panel line difference signals (LD) and their applications are the same.

On the other hand, panel portions of the line difference are increased to 7 μsec for overlapping, expansion then produces 1.3 MHz in 17.5 μsec. When line time in NTSC is doubled, two LD panel signals must be sent requiring quadrature modulation and double sideband applications resulting in 2.6 MHz in 17.5 μsec. The Yh (high resolution) component then becomes 2.8 MHz in 52 μsec. The total ''budget'' requirement now amounts to 2.8 MHz in 126.2 μsec, allowing 200 kHz for sound at 440 kbits/sec during frequency multiplexing.

The three augmented channel versions are illustrated in the composite figure supplied by Philips in Fig. 3-16, which we took the liberty of collecting and ''compressing.'' They show more of this interesting and innovative engineering effort by designers based in Briarcliff Manor, NY. Philips is already offering public demonstrations. The rest should be known later in 1990, depending on ATTC test times and reported results.

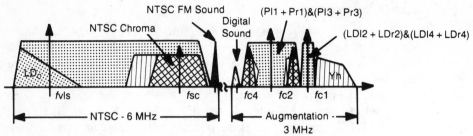

"6 + 3" MHz Version Analog Frequency Multiplex HDNTSC System.

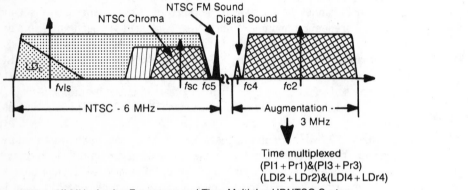

"6 + 3" MHz Analog Frequency and Time Multiplex HDNTSC System.

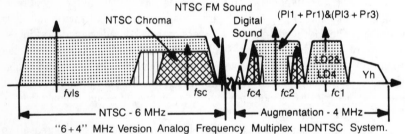

"6 + 4" MHz Version Analog Frequency Multiplex HDNTSC System.

Fig. 3-16. NTSC and augmented spectrums for HDTV as proposed by Philips (courtesy North American Philips).

PRODUCTION SERVICES' THE FOUR GENESYS TECHNOLOGIES

For pure mental stimulation, The Genesys Technologies by Production Services of Tucson, Arizona became one of the considerable surprises at the 100 + engineer show-and-tell November 1988 Advanced Television System five-day session in Springfield, Virginia.

Already in partial hardware, the Genesys name probably derives from the Greek word genesis, meaning birth or origin. Transparent (noninterfering) to AM, FM, or PM, according to Executive Vice-President Richard Gerdes, the complete system consists of four individual technologies that deliver high definition television signals at full bandwidth and yet remain *completely compatible* with NTSC on the *same* channel.

The four components are listed as:

1. Waveform carrier modulation (WM).
2. A/D/A analog to digital to analog conversion.
3. Digital bit compression (BITCOM).
4. Waveform modulation detection (ALLMOD).

System output bandwidth for RGB is projected at 30 MHz, 10 MHz each for red, blue, and green signals, and an audio with 16-bit dynamic range "approaching" 20 kbits and 20 kHz. Y-IQ outputs are also possible, with much of the receiver system contained within a 40-pin LSI chip. Basis for the entire project revolves about modulation added to existing transmissions that would not interfere with existing NTSC signals already there.

In this application, waveform modulation results from altering the shape of the carrier, since sine wave changes produce extra spectral energy, regardless of the source. Were amplitudes modified, only phase or frequency would generate sidebands but no harmonics. Nonsinusoidal waveshapes have harmonics and no sidebands. But continuously altering the carrier produces both sidebands and harmonics, and either can be selected for fully detectable modulation. Here "inflections" that appear as V-shaped "blips" on waveform slopes cause these carrier changes, and successful applications of waveform modulation will not excite AM, FM or PM (amplitude, frequency, or phase) detectors.

With intelligence modulated on each slope, sampling rates of $2 \times$ the carrier frequency are required, and eight analog levels or positions occupy each inflection. Since there are three digits/inflection, six bits transmit each carrier cycle. AM/PM transparencies are referred to the 3.58 MHz color subcarrier, AM carries color saturation and PM delivers phase or hue usually associated with tint. In stereo, the example continues, AM transmits in the left channel and PM in the L-R right. Waveform modulation can be executed by mechanical, diodes, or switches, and summing fundamental sine waves with especially phased third-harmonic sine waves is a fourth method. Sidebands about the fundamental and a dominant third harmonic are visible in spectrum plots.

ALLMOD Detector

Since modulation inflections occur along sinusoidal slopes, an ALLMOD detector was designed to subtract the carrier from a pure phase-locked sine wave generated from an LC tuned circuit and a variable resistor that establishes "Q" when driven from the input. Either a transistor or differential amplifier subtracts the input from the reference, and the difference becomes "the shape difference" resulting from waveform modulation. (See Fig. 3-17.)

When the tuned circuit's Q is low, amplitudes and phases of the reference sine wave change with AM/FM/PM modulations. Should the circuit Q go high, AM/FM/PM changes are not followed by the tuned circuit and the difference output is either AM, FM, or PM. These same signals can be separated from

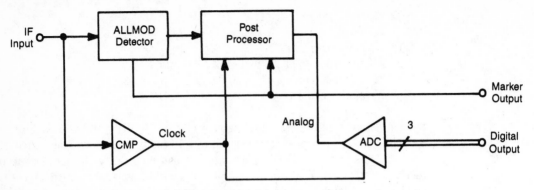

Fig. 3-17. The ALLMOD detector following the i-f input (courtesy Production Services, Inc.).

the WP detector by further processing in a sample and hold circuit collecting information from the WP output. Mr. Gerdes explains that by "altering the timing of the sampling, WP or conventional modulations can be distinguished." Later, it is expected that 4-5 bit values may be assigned each half cycle instead of 3 bits to either double or quadruple resolution.

A/D/A Conversions

Analog to digital conversions are accomplished by changing analog information to binary bits, outputting these bits either serially or in parallel. But an analog number has many possible values, and analog domain information converted to digital domain information is called quantizing and assigns each small analog variation some equivalent value (Fig. 3-18) in digital code. More bits means better analog resolution, but high bit rates frequently require fast parallel bit readouts. Delta modulators with serial outputs often generate errors, and are considered slow, therefore modifications were and are desirable. The Genesys system, consequently does the following:

- Modifies the Delta modulator for better response to both speed and differences in step sizes.
- The usual serial Delta output is combined with a standard parallel A/D converter for faster output as well as high resolution.
- The 3-digit modulation inflections are also monitored by a circuit that detects changes to generate a marker signal. And this results in multiplexer control to switch between binary bits and Delta Modulator output bits already combined in groups of three, as well as a receiver marker signal identifying most significant or least significant bits.
- A time constant has now been "assigned" to the size/difference of the Delta slope that's dependent on Delta error value and is proportional.
- Results are that the Delta modulation has no quantizing values and therefore produces, via the D/A section, an infinite resolution analog output with three digital bits.

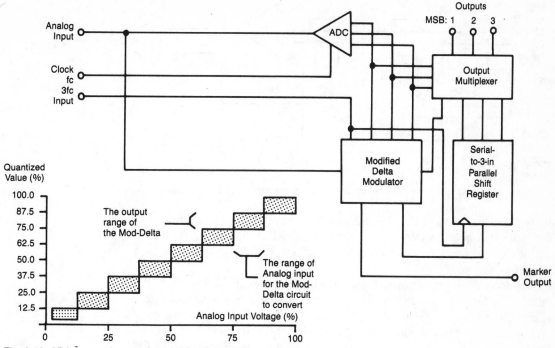

Fig. 3-18. ADA conversions and quantizing values (courtesy Production Services, Inc.).

Mr. Gerdes states the ADA synchronous conversion process retains a sampling frequency requirement $3\times$ higher than competing conventional systems. The bits, he continues, are not binary-related and can be processed through standard logic shift registers and stored in RAM (random access memory).

Digital Bit Compression

Shortened to BITCOM, we are reminded that at least 8 bits (2^3 in binary notation) are required for luminance and more for color; consequently, the need for bit compression in high resolution systems. CD players must have as many as 16 bits for maximum sound. The ear, however, can hear low frequency changes better than high frequency changes. In video, stationary objects have considerably more perceived detail than those that move—all identified as dynamic resolution. And while some bit compression systems use temporal filtering for bit reduction, memory is required for image reproduction and often generates motion artifacts. Audio usually warrants different treatment. BIT-COM, the Genesys system reports, represents real-time compression without memory and needs no image reconstruction (see Fig. 3-19).

The first three most significant bits (MSB) pass through from input to output, while less significant bits (LSB) are converted into groups of three bits, replicating actions of the A/D/A least significant bits, making BITCOM identical

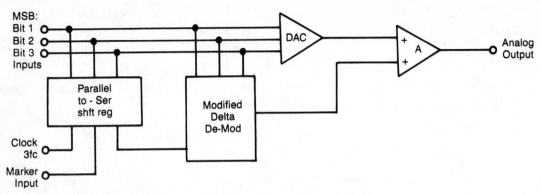

Fig. 3-19. Analog outputs from D/A conversions by the three most significant bits (courtesy Production Services, Inc.).

to A/D/A. This means that a 16-bit audio word compressed to 3 bits can directly drive the DAC in the A/D/A converter, and the 3 ADC bits of the A/D/A converter can also directly drive decompression in the BITCOM compressor for a linear bit-wide word.

In decompression, initial bits pass from input to output undisturbed, but additional information accumulated in 3-bit groups collect in a parallel-to-serial shift register. A serial output bit controls an inverse circuit to that of the compressor. The combined output then includes both least and most significant bits for a full-frequency and bandpass signal. The Genesys technology claims full compatibility with ADA bits, 3-bit compatibility with Waveform Modulation, a fast 3-bit transfer for best resolution audio/video, and the non-binary LSB processing permits full values in faster times.

Applications and Operations

Genesys reportedly is a spectrally efficient system transmitting 6 bits/carrier cycle. Requiring an IF exciter transmitter, it produces an effective bandwidth expansion factor of up to 10.4 and individual passbands of 10 MHz for RGB outputs—a total of 30 MHz. Two or four channels may offer compact disk (CD) sound, there is complete NTSC compatibility with no obsolescence of either today's receivers or transmitters, and it operates within each existing 6 MHz FCC-assigned VHF and UHF channel (Fig. 3-20).

While high frequency video information is bandlimited to 3 binary bits, slowly changing details offer resolutions up to 20 bits, and digital/analog converters maintain signal bandwidths unlike conventional binary converters. Should carrier modulation not be required, existing digital Tx systems can use regular digital transmitters with ADA and/or BITCOM. WM and ADA offer combined bandwidth compression factors of 10.48, and at standard NTSC video spectrum, the bandwidth of the system amounts to 44 MHz.

At the beginning of each line, a series of markers and an "illegal" state appears prior to setup bit transmissions for a calibration check. The entire

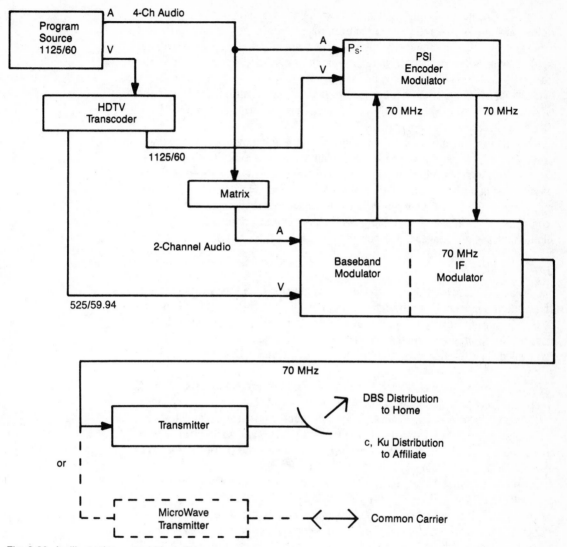

Fig. 3-20. An illustrative system block diagram (courtesy Production Services, Inc.).

frame is then integrated for additional calibration accuracy, ensuring correction of any differential phase distortion at the receiver prior to WM signal demodulation.

Audio transmits as digital bits among the video bits. Four channels require "four batches" of three bits each at a 67.5 kHz rate. Each channel may be muted with program source control bits, and the receiver can also respond to source control at 0 dB reference.

HDTV video inputs are at RGB, although the system is not dependent on any one scanning or aspect ratio format and simply transmits, compresses, and expands its input, being wholly compatible with both 4:3 and 16:9 HDTV aspect ratios. Signal-to-noise ratios are much improved due to processing more and less significant bits at identical amplitudes. And since this is a fully compatible, single channel system, there is no requirement for a second transmitter nor further UHF or VHF channel assignments.

Should 1990 hardware testing prove-out the successful operation of the Genesys four technologies, HDTV could be upon us in fairly short order. If not, then a two-step process appears inevitable.

We might also add that full system hardware implementation could well prove a number of skeptics wrong in assuming that such a radically different system can't operate successfully. For our thoughts, we are fascinated with the possibilities of both the transmission and signal detection proposals. There may be more Greek idioms to pursue than just Genesys, for in today's world of electronic marvels, virtually "anything goes." Unfortunately, some of the "proof" mathematics is not available just now.

JAPAN'S MUSE

Realizing that North American stakes are enormous in this highly competitive contest for HDTV approval and system installation throughout at least the U.S. and Canada, every effort has been made by NHK, the Japanese Broadcasting Corporation to argue their case before U.S. and Canadian engineers who will deliver the final report of the Advanced Television Systems Committee to the FCC sometime in 1991.

The Japanese have been working on their main FM project for more than 10 years and, therefore, have considerable experience in a basic system over most U.S. manufacturers and designers. However, theirs is primarily a satellite-delivered carrier with FM bandwidth of either 9 or 12 MHz, an aspect ratio of 16:9, 4-channel audio, and 1020 lines of resolution. But since the carrier is *frequency modulated,* it is not compatible with NTSC and therefore ineligible for terrestrial consideration. Further, since the MAC (multiple analog component) system advocated by Scientific Atlanta is already operating commercially nationwide, MUSE, the multiple sub-Nyquist sampling encoding scheme, could have difficulty even breaking into satellite transmissions on other than an experimental basis. Temporarily, at least, these are the two qualified or semi-qualified satellite transmission methods available for HDTV that could now actually go on the various transponders at both C and Ku bands. What develops around the 1990s might be something entirely different, but at least there are a pair of wideband audio/video vehicles awaiting someone's uplink with specialized receivers available on earth. Otherwise, only time and electronic ingenuity will tell.

The System

In Greek legend, MUSE described any of the nine nymphs or inferior divinities, said to be young, beautiful, and modest virgins, who dutifully promoted fine and liberal arts. They were Zeus' daughters who made sweet music together.

Unlike Greek mythology, the Japanese already have their 9/12 MHz bandwidth satellite systems in firm hardware, and another three different systems developing in computer simulation for a possible U.S. market. As explained to the Advanced Television Systems Committee by Taiji Nishizawa, Director, Advanced Television Systems Research Division, NHK, NTSC-MUSE 6, 9, and Narrow MUSE would have respective bandwidths in MHz, producing 750 lines of horizontal resolution for the 6-9 MHz systems and 1010 for Narrow MUSE. All systems would offer aspect ratios of 16:9, with either 4-channel or 2-channel digital audio, and all are based on the 1125/60 international standard. This is reduced to 750 in the MUSE-6 and MUSE-9 systems, lessening vertical resolution but comparing favorably with 1125 lines subject to a Kell factor of 0.65. When decoded, however, the 1125-line count is restored at the receiver.

In reducing signal bandwidths, multiple sub-Nyquist sampling occurs for still picture elements consisting of both field and frame offset sampling. Moving elements require only line offset sampling. At the decoder, both interfield and interframe interpolation takes place for stationary images, while only intrafield interpolation is required for moving portions. As for audio in either 2 or 4 channels, a scheme named DPCM Audio Near Instantaneous Compression and Expanding—DANCE—is used.

Narrow MUSE

The system probably pushed hardest by the Japanese is Narrow MUSE (Fig. 3-21). They claim channel transmissions are within NTSC's 6 MHz and there are no restrictions on NTSC compatibility since this is a standalone system. Four digital audio channels are transmittable, signal processing is simple, and the system remains compatible with parent MUSE. Conversely, 6 MHz MUSE is NTSC-compatible and will transmit two digital audio channels, while 9 MHz NTSC-compatible MUSE requires a 3 MHz augmentation channel to improve motion resolution. Experimental equipment under construction for demonstrations became available in 1989, according to the Japanese, who are working hard to maintain any sort of lead over U.S. firms who operate individually and not in a government/industry/marketing consortium.

Since Narrow MUSE has identical encoding and decoding processes except for scanning line conversions as wideband MUSE, only interface encoder/decoder adapters are required to couple the two systems. For encoding, separate luminance and two chroma signals are sampled at 40.095 MHz and 13.365 MHz apiece before entry into a 1125/750-line scan converter. Here, scan lines are interpolated and 750 lines extracted. During stationary scenes, lines from the immediate and prior two fields are interpolated; but for movement, only the existing (present) field is involved. Line converter output after filtering passes directly into the MUSE encoder.

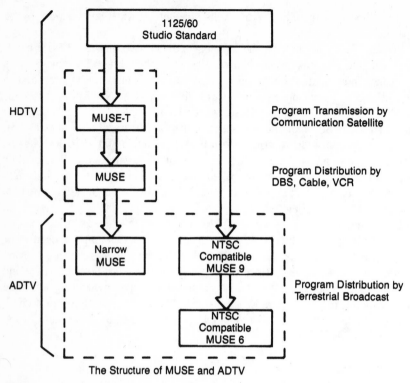

The Structure of MUSE and ADTV

Fig. 3-21. Block diagram illustrating relationships among the MUSE systems.

The interface adapter then converts the 750-line information with 1188 picture elements to a 1440-pixel signal with 1125 lines for standard MUSE by the addition of extra picture elements. Regular encoding may now proceed (Fig. 3-22).

During picture motion, a diamond prefilter engages for line offset subsampling at 48.6 MHz. Another diamond prefilter is also used for motionless scenes with field offset subsampling at 24.3 MHz. Higher frequencies are rejected by a 12 MHz low-pass filter, with sampling now converted to 32.4 MHz. The moving and stationary pixel groups are combined according to the amount of motion while being frame offset and subsampled once more, but now at 16.2 MHz, Narrow MUSE has a video bandwidth of 4.86 MHz (Fig. 3-23).

In decoding, Narrow MUSE inserts into a MUSE format frame to reproduce the 1125/60 format video as well as audio.

Audio Encoding/Decoding

Audio in the MUSE systems is either 2 or 4 channels, but wholly digitized. This encoded information is multiplexed in the VBI vertical blanking interval

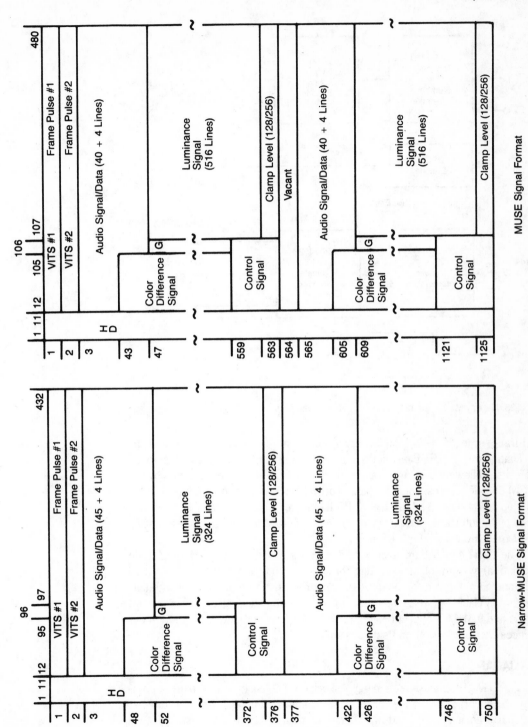

Fig. 3-22. The two main Japanese signal formats (courtesy NHK Japan).

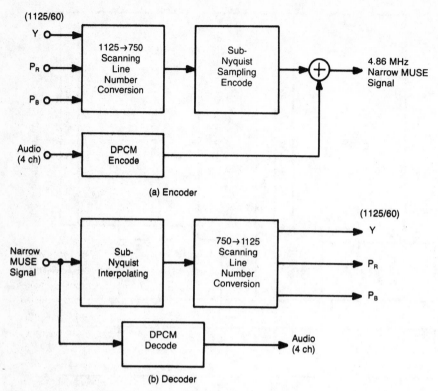

Fig. 3-23. Narrow MUSE encoding and decoding examples (courtesy NHK Japan).

field/frame scans. The 4 channels are specified at 15 kHz bandwidth, and the two channels at 20 kHz bandwidth. The first is sampled at 32 kHz and digitized to 15 bits. Following this, 3-bit range and 8-bit DPCM data are extracted by the DPCM encoder. The second is sampled at 48 kHz, digitized into 16 bits, DPCM encoded and compressed to 11 bits. Both bits are serialized, error codes introduced by bit interlace, and binary becomes converted to ternary code following time/axis compression for VBI insertion.

The NHK consortium is also working on a possible VSB-AM (amplitude modulation) system for simulcasting that would produce a bandwidth of 5.946 MHz and SAW-filtered for Nyquist characteristics required. Additional details were not available before press time. Top and bottom masking is used in the two 6 MHz and 9 MHz NTSC compatible systems to prevent picture resolution differences between center and side pictures.

BTA JAPAN

The BTA (Broadcast Technology Association) consists of 43 Japanese companies and corporations identified with manufacturing and broadcasting, is headquartered in Tokyo, and was established in 1985. In their study of an extended definition system (EDTV-1) for terrestrial television, 25 systems were proposed, 10 of the 25 survived, and 8 can be demonstrated.

In Phase I evaluations, the 8 systems were judged on 18 stationary pictures, three of which were sourced from the ITE digital standing chart with interlaced camera, six were scan-converted from a progressive scanning camera, and nine from a downconverted Hi-Vision camera, with contents ranging from women models to gardens, a weather forecast, buildings, computer graphics, dolls, fruit, and even a candy box.

Following this, compatibility tests were undertaken to discover whether these systems created additional artifacts or interference with NTSC-type receivers or video cassette recorders. Seven of the eight reportedly passed. Afterwards, it was shown that a combination of 3-dimensional luma/chroma separation, 3-dimensional receiver scan conversion, and higher resolution signals at the transmitter were of considerable help in improving picture quality (Fig. 3-24).

For Phase II, in addition to the triple combination described above and identified by the Japanese as EDF, precompensation of saturated colors in "quasi-constant luma signal processing" called Y3, and adaptive high frequency component emphasis in luminance (S1), was also found helpful.

Tests in Phase II were carried out in the laboratory to ascertain picture quality improvement and receiver/VCR compatibility, as well as with broadcast equipment, and in the field with terrestrial transmissions to check EDTV effects with noise, secondary images, and other possible problems, including signal degradation. At the laboratory, the same stationary pictures were recorded in both NTSC and EDTV formats with a 1-inch VTR, and then tapes were distributed to 12 TV manufacturers to check compatibility with their own television sets and recorders. A total of 54 receivers and 39 VCRs were involved. Compatibility was generally good. Over-air transmissions also posed no problems, and terrestrial tests did not increase audio buzz perceptibly. Finally, EDTV test signals were demodulated and recorded on the same 1-inch VDTR. In good reception areas, results were very good, but in fringe areas there was also some reduction in thermal and pulse noise compared to NTSC. In short, "no problems," according to BTA.

In the Japanese report there was also mention of a ghost-cancelling reference signal at the transmitter, correction of dark picture details, and highsat color pre-compensation. At the receiver, ghost cancellation was specified, as well as special Y/C separation and progressive scanning. A block diagram is shown in Fig. 3-25 for the receiver.

Second generation EDTV systems will also feature wide aspect ratios, higher resolution, and full fidelity sound. And in HDTV, there will also be a wide aspect ratio planned, twice the usual NTSC picture resolution, and Hi-Fi sound.

The prime information available on BTA is primarily testing rather than the introduction of new systems that other media such as MUSE advocates are

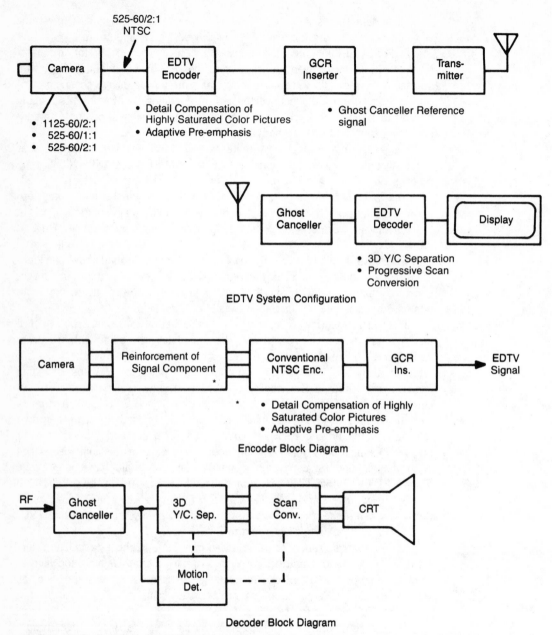

Fig. 3-24. Block diagrams of EDTV-1 system configuration (courtesy BTA Japan).

expected to generate. With all the east-west development and testing under-way, however, some remarkable developments in both terrestrial and satellite Tx/Rx are expected.

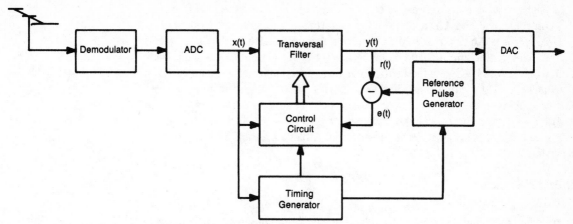

Fig. 3-25. Block of ghost cancellation receiver (courtesy BTA Japan).

FURTHER JAPANESE VIDEO DEVELOPMENTS

There are some additional solid state and video developments that either connect directly with HDTV or supplement available video information, filling in some of the holes that might not otherwise surface. And in pursuit of this knowledge, we will borrow both information and some diagrams from the IEEE Consumer Electronics Transactions of late 1988. This is done so that today's concepts appearing in hardware several years down the road will be covered, perhaps not in precisely the identical form described in this chapter, but with enough similarity to recognize the principles and origins. In consumer products audio/video offerings, the Japanese have a virtual lock on the market, especially in video cassette recorders, players, cameras, camcorders, radios, and the like. But in television, the U.S. still has Thompson/RCA, Zenith, Philips (Magnavox, Sylvania, and Philco), and many monitor and computer manufacturers, as well as video game sources. So there's still healthy competition in the consumer market despite substantial Oriental inroads in many categories.

Field Storage

In this new product, employees of Texas Instruments Japan, Ltd. showcased a brand new video field store system with single 1-megabit memory having 6-bit resolution and 3-field store-sampled conversion (Fsc). The field memory has 4-bit high speed read/write data ports that permit special effects such as still, strobe, and search, which also may be applied to video disks recorded in the constant linear velocity (CLV) mode.

In executing still picture format, a 0.5-line skew is avoided by storing 263 lines of a single field, then reading out 263 and 262 lines alternately for the usual 525-line frame, with an average field frequency of 60 Hz, and a frame rate of 30 Hz.

The system actually consists of four integrated circuits, one field memory, a controller, and one A/D and one D/A converter. Its composite video input A/D-converts to 6-bit data with 3Fsc sampling. A 6-4 converter reduces the 6 bits to 4 to be written into the field memory, and a 4-6 converter changes 4 bits back to 6 bits for the DAC, and the digital-to-analog operation returns the information to analog. At the same time, V/H sync pulses are separated from composite video. Together with vertical sync, two phase-locked loop oscillators generate $4.5 \times$ Fsc from the color burst and memory controls for clock, write-enable, reset-write, serial-read clock and reset-read. A block diagram of the field store system is illustrated in Fig. 3-26.

EDTV Studies

While MUSE remains the prime Japanese HDTV system, their engineers are also continuing work on intermediate enhanced definition systems some of which are already in existence, including 16:9 aspect ratios and other significant improvements. Don't attempt to estimate a working retail price on these wide-angle picture tubes, however, since non-production units would probably exceed $2,000. Based on a $1400 price tag for 4:3 35-inchers, we're probably not far off the mark just now. By 1991, all this may change, but probably not substantially. Big tubes make big pictures but they also require rugged price tags.

In the meantime, extensive work continues with EDTV in many laboratories in Japan and abroad, for EDTV will probably become the first considerable expansion of NTSC before actual HDTV reaches the market. To this end, television engineers of the Nippon Television Network have been attacking the problem and are making their usual progress.

Initial considerations were:

■ Horizontal luminance resolution
■ Vertical luminance resolution
■ Chroma resolution
■ Elimination of crosscolor and crossluminance
■ Improved signal-to-noise (S/N) ratios
■ Better picture definition

It was found that progressive scanning promoted improvement, special gamma corrections were instituted at primary RGB color stages and matrixed for the usual Y luminance and I and Q color information. I and Q was then band limited to 1.5 MHz for I and 500 kHz for Q, then NTSC encoded, although this restricts high-band luminance information, decreasing resolution. By applying adaptive emphasis, extended luminance is attained allowing pre-emphasis for added resolution.

Comb filter three-dimension Y/C separation for still pictures effectively rejects crosscolor and crossluminance, but moving images respond to two-dimension chroma/luma separation. These engineers say an ideal separator

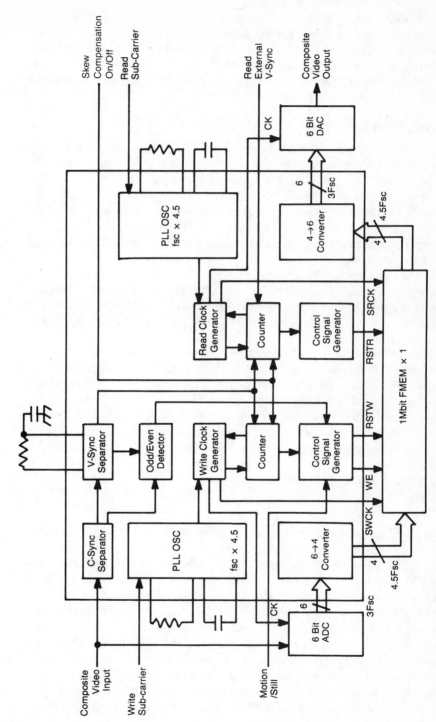

Fig. 3-26. Enhanced definition TV system compatible with NTSC (courtesy IEEE Consumer Electronics and NHK).

would offer a dual comb filter system, depending on the scene. The addition of progressive scanning removes visible scanning lines and removes all line flicker, improves vertical resolution, and delivers better picture resolution. The color encoder was then made to switch bandwidths in the I portion from 1.5 to 2 MHz upon command, and there is a de-emphasis circuit in the receiver. Figure 3-27 shows the system configuration.

Fukinuki/Hitachi

New research is reported by Dr. Takahiko Fukinuki and associates of Hitachi who claim development of a fully compatible EDTV system that improves both luma and chroma by introducing an additional subcarrier. A proponent of the Fukinuki "hole," he explained this in terms of the 3-D frequency spectrum. Color information is placed in the second and fourth quadrants of the traditional circle as luminance (Y) modulation is limited to avoid aliasing in either vertical or horizontal regions, the first and third quadrants being little used (Fig. 3-28).

The FUNCE—or fully compatible EDTV development—extends horizontal resolution by placing higher resolution information in the Fukinuki hole and improving vertical resolution by progressive/interlace scan conversion with 3-D filtering, interlace, and pro-scan displays. Horizontal resolution responds to new subcarrier modulation.

Color improvements include Gamma compensation correction and possibly swapping I and Q bandwidths of 1.5 MHz and 500 kHz, respectively, on odd and even fields. But according to the published description, upper sideband Q was enlarged to 1.5 MHz and frequencies above 4.2 MHz are inserted into the existing frequency band by modulating the subcarrier with luma and chroma extended information. There are, apparently other ways of extending the Q bandwidth also because of the original 500 kHz spectrum spread. Apparently the 1.5 MHz I information remains in place with both luma and chroma extensions. No maximum Y resolution parameters were given.

Wide Screen EDTV

Five Japanese engineers who work for Matsushita in Japan are responsible for this NTSC-compatible, wide-screen television system with some very interesting innovations. Still in the laboratory and possibly for later production, the QUAM (for quadrature amplitude modulation of the video carrier) with an inverse Nyquist filter is actually multiplexing the added signal without disturbing NTSC. This is done at the transmitter, and when QUAM reaches the receiver an added Nyquist filter shapes the multiplexed portion into double sidebands, reducing power so that it produces no crosstalk with PLL (phase-locked loop) synchronous detectors and only "a small amount" with diode envelope detectors—that is, for standard-type NTSC receivers. In the wideband system, the multiplexed information is detected by another PLL synchronous detector following band limiting and main NTSC becomes double sideband. Sidepanels

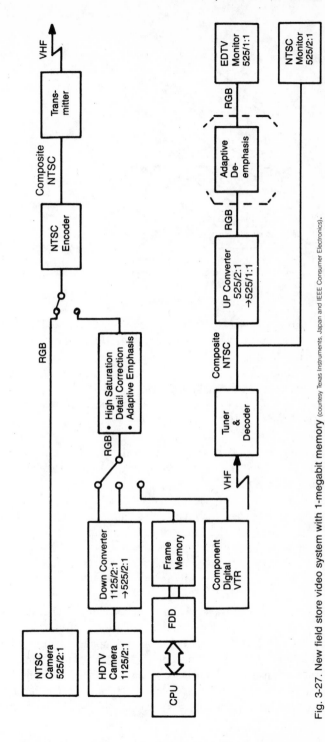

Fig. 3-27. New field store video system with 1-megabit memory (courtesy Texas Instruments, Japan and IEEE Consumer Electronics).

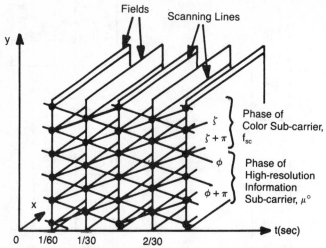

Fig. 3-28. The Fukinuki "hole" as seen in three dimensions with scan lines and subcarrier phase relationships (courtesy Hitachi and IEEE Consumer Electronics).

immediately expand the aspect ratio to a wide screen picture following subchannel detection. The new aspect ratio then becomes 16:9, which seems to be the accepted standard for eventual HDTV.

In implementing the system, a 16:9 aspect ratio camera with a passband of 5.6 MHz is used in the studio and the usual 3.579545 MHz chroma subcarrier. Switchers and synchronizers convey signals to the transmitter where the extended NTSC information has to be decoded into component signals with frame memories. This provides a conventional 4:3 NTSC aspect ratio plus the sidepanels that are delivered in the quadrature channel. Becoming a compatible NTSC signal, it can now be received by regular as well as extended definition and aspect ratio receivers.

Secondary images or Nyquist filter mismatches between transmitter and receiver, however, had to be overcome. The solution was to transmit the multiplex signal in alternate polarities either line/line, field/field, or dual dimensional waveforms can be developed with equalization for both main and multiplexed signals. By careful control, both ghost and channel crosstalk can be improved. At the hardware stage, this second method was used with an equalizing range of some 36 μsec.

Block diagrams of the newest encoder and decoder are illustrated in Fig. 3-29. In the encoder, red, blue, and green inputs are matrixed to luminance and color signals I and Q. Low-frequency sidepanel components are 5× time compressed and multiplexed to the front porch of the horizontal blanking period. High-frequency sidepanel components are time expanded ×4, and both luminance and the I color information are center expanded, according to the diagram. Sidepanel luminance then passes through a high-pass filter, is inverted,

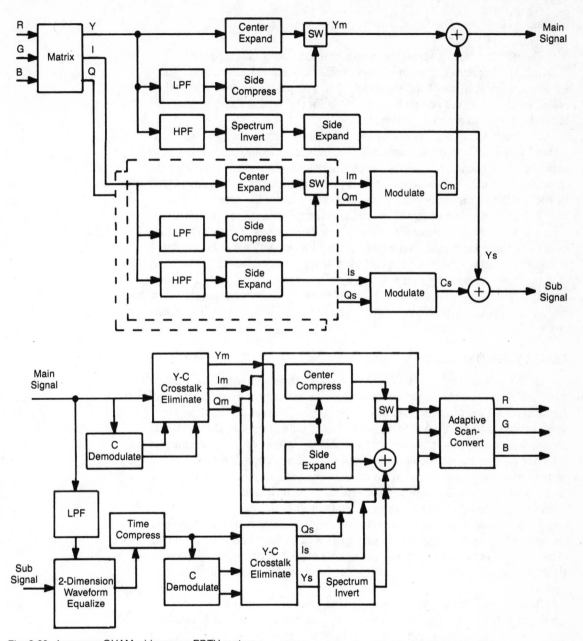

Fig. 3-29. Japanese QUAM widescreen EDTV system (courtesy IEEE Consumer Electronics and Matsushita Electrical Industrial Co., LTD.).

then side expanded and added to the subsignal. I information is similarly compressed and expanded following passage through high-pass and low-pass filters, and I and Q are modulated together with luminance side expansion for the subside signal. The main signal then receives center expanded luminance in addition to low-pass filtering and center-expanded I, along with Q from the modulator.

The decoder receives both the main signal and subsignals, a portion of the main signal being routed through a low-pass filter into a 2-dimensional waveform equalizer, there joined with the subsignal. Time compression immediately takes place, color is detected and then combined in the Y/C crosstalk eliminator, followed by spectrum inversion of luminance.

The main signal also has color demodulation, with luminance and color joined in the Y/C crosstalk eliminator. Luma center compression and side expansion then occurs, followed by subchannel recombination and RGB outputs through an adaptive scan converter. In this process, the main signal is expanded by $5\times$ and the subsignal time compressed by $4\times$, exactly the reverse of encoding. The final step is an adaptive scanning converter using a motion detector with frame memories. Vertical resolution now becomes 480 lines. The engineers say time compression and expansion takes place with linear interpolation and filters. Hard details here are somewhat skimpy, but so was the translation and, perhaps, some of the interpolation. Quadrature amplitude modulation may appear attractive in the laboratory, but implementation for full production and Tx/Rx could produce problems if not executed precisely and probably expensively at these frequencies.

QUANTV TECHNOLOGY

QuanTV is not truly an HDTV system but a process designed to quantize various Y, I, Q luma-chroma levels for both digital and many analog video transmissions, in addition to dithering, which is expressed by the dictionary as "nervous excitement or shaking." Here, dithering results in pel-to-pel fluctuations among consecutive gray-scale levels that's said to "preserve" fine shading and low contrast picture edges otherwise overlooked. This dithering varies from pel to pel, line to line, and frame to frame as it operates to contain both picture detail and resolution. As such they are repetitive perturbation signals extracted from memory or processed with logic circuits or shift registers.

At the same time, Nyquist luma and chroma information is quantized coarsely following dithering, and image details with sufficient contrast to include single or multiple luminance quantization are sampled during every element of the raster in each frame and eventually reproduced with "full spatial and motion resolution."

In color processing, any variable chroma "vector" is dithered and quantized and only specific chroma "states" are transmitted. In digital transmissions, QuanTV reduces both bits/sample and bandwidth; but in analog operations, S/N and interference resistance are emphasized rather than bandwidth reductions.

Coarsely quantized Nyquist samples in this example can be scrambled at quantum levels by cryptographic keying.

Conventional TV receivers accept Y and IQ information like "corresponding" video and there is little or no loss of picture quality, according to Quanticon. Y, I, and Q could also be regenerated in a modified home receiver to remove noise, interference, secondary images, cross modulation, and various distortions.

In published statements, Quanticon declares it "does not propose to construct system hardware . . . but will cooperate with other proponents." Licenses under issued and pending Quanticon patents are available to all qualified parties on a fair and equitable basis.

IMPROVED DEFINITION AT THE TRANSMITTER (HRS)

This is another system with interesting possibilities that requires either NO changes in existing television receivers or simply small ones to some tape recorders consisting primarily of modifying several counters in the genlock (locking to a master generator) circuit for better sync. Primarily, these are 1 in. tape equipment that derives sync by counting down from the 3.58 MHz chroma subcarrier.

Under current cable and terrestrial broadcast testing, the "head end" system is designed to eliminate dot and chroma crawl and cross color in ordinary TV receivers after slight transmitter modifications. Named Chroma Crawl Free™ or CCF™, the High Resolution Sciences product re-synchronizes and encodes TV pictures at RGB and encoder levels or converts standard NTSC with CCF digital frame store.

As many of you probably already know, luma/color in NTSC is interleaved at half the line scanning frequency in odd multiples. Lines contain 227.5 cycles and each frame has 119,437.5 subcarrier periods, plus every field ends with the subcarrier 90° out of phase with the previous field. Therefore, in each consecutive field the color subcarrier position is one line above. This, then, is the reason for the chroma/dot crawl effect that often appears at the edges of bright colors or color patterns where dots seem to be moving slowly upward. In cross color, receiver demodulation mixes luma and color in certain fine detail or sharp black/white transitions moving some colors upscreen or showing spurious shimmering color in luma image portions.

In CCF, the 227.5 cycles of subcarrier/line are modified to 227 cycles or an increase in horizontal and vertical frequency of 2%, resulting in all lines in the same field maintaining the same (and unchanged) subcarrier phase. The half cycle, however, remains in the last line of each field so that the next field produces subcarrier 180° out of phase. Consequently, fields 1/3 have identical subcarrier phase in every line, while fields 2/4 are in opposing phase to 1/3. This sync pattern eliminates chroma crawl, according to High Resolution Sciences and, since same-frame adjacent lines have out of phase subcarrier, cross color is cancelled also. A block diagram furnished by High Resolution Sciences (Fig. 3-30) clearly illustrates the electronics required to operate the system at either

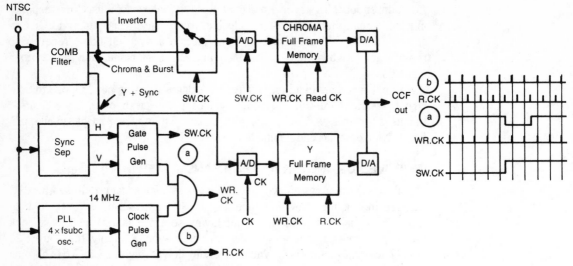

Fig. 3-30. Block diagram illustrating the HRS system (courtesy High Resolution Sciences).

head end or broadcast transmitter. The 3.579,545 MHz traditional color sub-carrier frequency remains the same. With some artifacts and coloration removed, this is an improved definition picture, but certainly not the HDTV that will eventually appear.

THE OSBORNE SYSTEM

Based on Patent 4,665,436, dated 5/12/87, Joseph Osborne and Cindy Seiffert have also offered a means of compressing digital data for both transmission and reception into narrow bandwidths using convolution or averaging techniques, then sampling and predicting data points, subtracting data from predicted values and transmitting data and difference information to a receiver with the same prediction algorithm. Transformed image and encoded difference signals are then applied to the predictor, subsampled, and added to the decoded difference signal, recovering raw data.

According to Osborne Associates, baseband video bandwidths are selectable, depending on transmissions within a single 6 MHz channel, two 6 MHz channels, or one 9 MHz channel. Said to be compatible with existing receivers, channels, and transmitters, the single channel offering can deliver full digital transmissions with QPSK (quadrature phase shift keying) modulation within a 70 MHz i-f, including CATV systems. Single channel satellite transmission using QPSK is also possible.

In single channel digital within a 6 MHz system, both NTSC and the error signal can be compressed within a 45M bit data rate over fiberoptic cable in a T3 format. In another version, NTSC would be transmitted en toto with an extra 3

MHz consisting of the error signal, which actually corrects any artifacts. Luminance spatial and temporal resolutions are said to be maintained, depending on channel and originating source.

Patent declaration states that the ''present'' invention is a method of reducing the rate of pixel changes in sampled video by local averaging and encoding the differences. Transformed signals depict pixel energy in one frame, representing the original image. Differences between pixels that are artificially smoothed along with those transformed are encoded for minimum word length. The receiver is said to recover virtually the entire ''raw'' input information at its output. Osborne claims optimum efficiency of digital data communications by reducing transmitted bits and transmission bandwidths.

Transmission times are constricted by not sending the complete raw waveform while an A/D converter repetitively samples the analog video signal level at a set rate. For instance, in a single frame of video image, the sampled signal appears as a succession of dots or pixels, said to accumulate to some 1.3×10^6 pixels/frame. Video compression uses a ''multilevel encoding scheme, including certain characteristics of difference transmission and characteristics of temporal and spatial compression, together with a minimum bit count data format.''

As you can see in the combined transmission and receiver diagrams (Fig. 3-31) A/D operations are followed by filtering and sampling, then a predictor circuit feeds a subtractor that supplies difference information for the following field selector and encoder. Meanwhile, the output of the filter and sampler is also transform coded, normalized, and then encoded, where it joins the subtracted information in the demultiplex for digital data output.

In the receiver below, this digital data is collected and multiplexed into two decoders. The top unit is inversely normalized and transformed, passed into a predictor, field selected, and joins the other decoded information from the multiplex in an adder. Interpolation then occurs and a digital to analog (D/A) conversion takes place for reconstitution of standard video. According to Osborne, transmissions can take place over wires, electromagnetic radiation or light, via satellites, etc.

As you may be aware, patent applications are not always subjects for lucid translation. This one, similar to one or two others, was a little difficult.

SCIENTIFIC ATLANTA

Like Japan's MUSE E and Philips' HDMAC-60, this is a purely satellite-delivered high definition television system. It is designed to operate by frequency modulation but within the multiplexed analog component (MAC) process originally developed in Great Britain by Drs. Wyndham and Lucas, which was refined in Canada and the U.S. by Scientific Atlanta. There are at least three versions extant of which B-MAC is now used in business communications by a number of commercial companies across America. As of 1989, HDB-MAC (as the system is called) was in hardware development, and initial demonstration

88

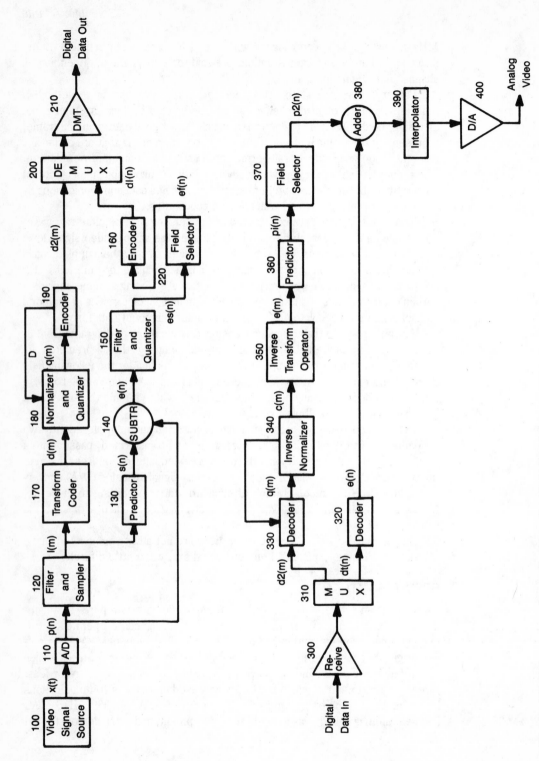

Fig. 3-31. The Osborne system transmit and receive block diagrams (courtesy Osborne Associates).

of conditional access was projected for the spring of 1989. Commercial availability could come that summer. Unlike terrestrial strictly-controlled single and dual 6 MHz RF channels, HDB-MAC is intended for a *video* passband exceeding 6 MHz and is compatible with all normal B-MAC decoders. With delivery to CATV head ends and ordinary broadcast stations, the format is easily transcoded for use by those two media.

As a feeder signal, the system produces 525-line sequential scan component pictures in 4:3 and/or 16:9 aspect ratios, with minimum artifacts or other picture problems, along with digital audio in six Dolby channels, plus data. Decoders possess both component and NTSC outputs, and all "services" (probably programming) are both addressable and encrypted. A single line of this multiplexed system is illustrated in Fig. 3-32. It shows 11 μsec, 17.5 μsec, and 35 μsec reserved for data, color, and luminance, with the data portion varying when scrambled. These three entities add up to the usual 63.5 μsec NTSC line scan time, including the horizontal blanking interval.

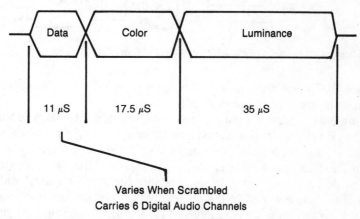

Fig. 3-32. Horizontal line multiplex format for HDB-MAC (courtesy Scientific Atlanta).

System Parameters

Parameters for the system include: 525 lines (vertical), 2:1 interlace, and a field frequency rate of 59.94 Hz; transmissions are separate luminance (Y) and color difference signals (R-Y and B-Y) in sequence; a transmit bandwidth of 6.3 MHz; luminance at 4.2 MHz; chroma bandwidth of 1.5 MHz; dual 4:3 and 16:9 aspect ratios; and a 17.5 MHz FM deviation in a 24 MHz passband transponder channel. A block diagram of the B-MAC chip set is shown in Fig. 3-33. Here you see the various controls, channel generator, random access memories (RAMs), and the usual L/R stereo sound and RGB or NTSC video outputs. Enhancements include doubling the line scanning frequency, increasing both vertical and horizontal resolution. Static vertical resolution is increased without generating motion artifacts, fieldstore adaptive scan conversion offers good results, and the motion-adaptive field counter is able to switch without introducing artifacts. In

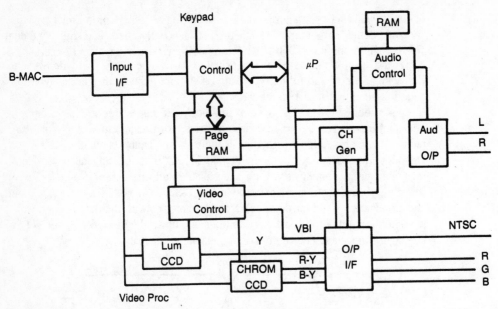

Fig. 3-33. Integrated circuits that make up the B-MAC system (courtesy Scientific Atlanta).

luminance HDTV spectrum folding, the signal reportedly measures some 7 MHz, and with time compression, 10 MHz.

Figure 3-34 illustrates luma/chroma processing and HDB-MAC decoder action. Nyquist characteristics approximate 1/2 sample, along with vertical filter interpolators, line decimators, filters and 3:2 or 3:1 time compression before reaching the Y/C/data multiplexer and then the 10.7 MHz skew-symmetric filter. The decoder then time-expands luminance 3:2 and chroma 3:1 with an inverse of the process used in the encoder. Static vertical line resolution is reported to be 480 lines, with dynamic at 240. The HDB luminance resolution chart shows both 735 and 945 lines of horizontal resolution. Colorimetry reportedly is the same as NTSC. R-Y and B-Y are said to issue from the decoder at 5 MHz and luminance at 18 MHz. For reception and CRT display, you will need an RGB processor.

Broadcast IDTV—Faroudja

California's Faroudja Research Enterprises, Inc. has proposed a SuperNTSC™ improved definition TV system that would occupy the same 6 MHz frequency allotment now serving the 69-channel U.S. VHF-UHF FCC-assigned broadcast medium plus CATV.

Based on the letterbox concept for NTSC compatible image and the usual 59.94 fields, Faroudja would maximize camera image pickup along with RGB and progressive scanning (1050 lines) and, by matrixing Y luminance and I-Q

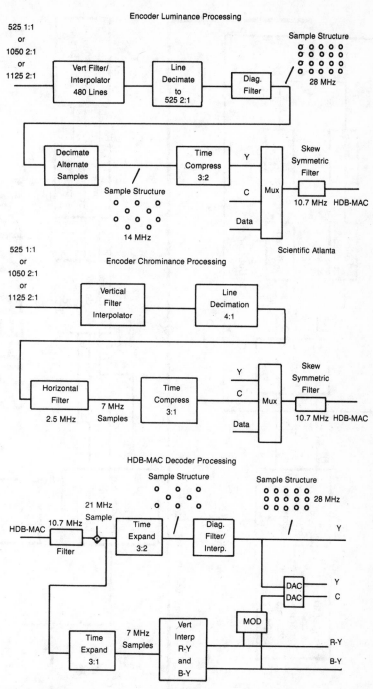

Fig. 3-34. General block diagrams of luma/chroma decoders in HDB-MAC signal processing for satellite distribution (courtesy Scientific Atlanta).

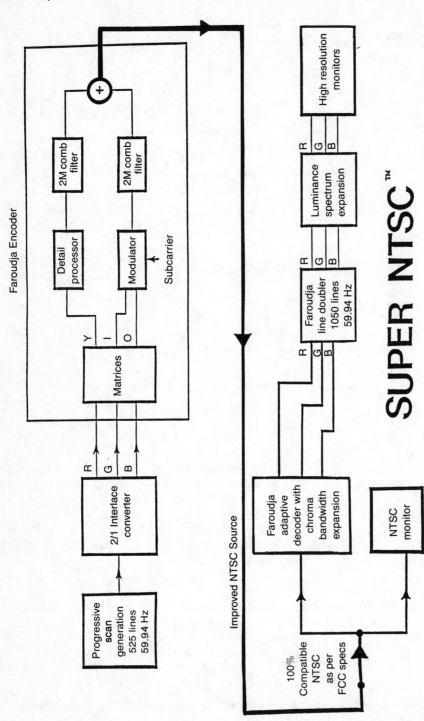

Fig. 3-35. Faroudja's improved definition super system for current NTSC broadcasting. (courtesy Faroudja Research Enterprises, Inc.)

color, followed by subcarrier modulation and detail processing, then restores all parameters to 100% compatible NTSC for both transmission and reception.

By so doing, Faroudja will deliver 400 scan lines per active field compared to 480 lines by competition. This means, however, the Faroudja system cannot accept a cropped picture or one that is underscanned so that full detail is presented equivalent to proponents using the 16:9 aspect ratio "except for a 2% loss on each side (of the scan) and a 2% gain top and bottom because of the difference in aspect ratio."

In decoding, Faroudja again uses adaptive 2H comb filters, two delay matches, quadrature color demodulation with subcarrier and an R-Y, B-Y delivery that's matrixed with Y luminance for Red, Green, and Blue outputs. Included in the overall system are both luminance and chroma bandwidth expansion. The Research Enterprise is still working on ghost and micro-reflection reducer, noise reducer, TV receiver IC cost reduction and end-to-end testing.

An assessment of ATV systems and technologies by SS Working Party 1 recognizes the improved picture quality of the system "without augmented information" and states SuperNTSC is complete and in full hardware. But says that since no additional information has been added to NTSC, "there is some question as to whether it should be classified as an ATV system, even though a quite noticeable enhancement in *perceived* resolution has been demonstrated." See Fig. 3-35.

4
HDTV Digital Sound and MTS

NOT ONLY WILL HDTV BRING BETTER VIDEO BUT IMMEASURABLY IMPROVED sound, rivaling audio from the now-famous digital disk. In addition it will not have to be transported by separate FM carrier, but probably be buried as digital bitstreams in one of the convenient blanking intervals. Add selective multiplexing, perhaps further enhancement and surround sound, and audio could well become a video-equal instead of its traditional second cousin. For even with the introduction of multichannel BTSC-dbx stereo/SAP sound broadcasting, we have yet to hear anything that competes head-to-head with discrete and/or stereo multiplex satellite sound. Now add all the luxuries of digital processing to already superior analog reproduction, and you have an audible medium that can offer the best music and voice imaginable. If all this can be accomplished without picture artifacts (disturbances), then HDTV can also become HDSTV, too. At the moment there is every indication it can, but nothing becomes final without thorough testing in laboratories and on the air.

In digital systems, audio must be treated as sampled data, often between 40 and 50 kHz musically and 8 to 10 kHz for voice due to the usual 20 kHz rolloff and the telephone's bandwidth limitation of 4 kHz. Audio treatments in various systems are extremely diverse, including mixing and equalization, signal conditioning, filtering, voice store, speech recognition, synthesis, vocoding, in addition to multiplication and accumulation. All deal with a few or a considerable number of speech samples which, according to engineers, can total as many as 10,000 samples/second. Fortunately, programmable digital signal processors (DSPs) are increasingly available with internal A/D and D/A converters that aid this sound processing immeasurably.

In our HDSTV (high definition TV-sound) category, two major companies are competing for the audio privilege. Both are recommended by HDTV video systems groups, since most of these groups will not furnish their own audio. Proponents Dolby and Digideck are located primarily in California. Their system descriptions should prove interesting. And since we have the Digideck information immediately at hand, their system will be described first.

DIGIDECK, INC.

This Mountain View, California, company proposes not one, but two high fidelity stereo systems designed to rival CD disk performance in the home. A pair of "strawman" examples are offered since data, available clock rates, and channel characteristics in other portions of the HDTV package are not yet specified sufficiently for adequate combination (Fig. 4-1).

One Digideck system is "direct." Operating at 650 kilobits/sec, it produces a 20 kHz stereo pair, while the other can be described as a "basic, plus augmentation" design that delivers a 15 kHz stereo pair at 475 kilobits. This second method may also be upgraded later to 20 kHz with an additional augmentation effect. Assumed are 16-bit pulse code modulation (PCM) inputs and outputs at typical CD sample rates of 44.0559 kbps.

Both systems have comparable similarities; especially an ADEX algorithm that stands for "adaptive differential entropy coding," and is said to be a special combination of adaptive processing, differential PCM and entropy coding. Extensive computer simulations have preceded realtime hardware development. There is also an error detection and correction scheme, and a process that masks audio segments lost through "unrecoverable" transmission dropouts or mistakes. Other portions of these audio systems are said to be relatively conventional, and include high quality 16-bit PCM digital audio reception, a digital transmission channel having independent random and burst errors, bit and timing recoveries, plus both frame sync and sample-rate clock, and 16-bit PCM

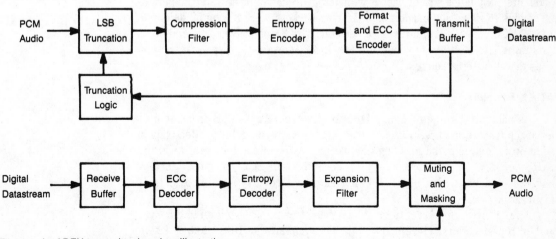

Fig. 4-1. An ADEX transmitter/receiver illustration (courtesy Digideck).

recovery back to analog. "Near CD fidelity" is predicted at some six bits/sample.

ADEX Processing

Assuming a 16-bit PCM analog conversion, the ADEX method will process this audio digitally and deliver 1470 samples at a 29.97 Hz frame rate to the decoder for D/A reconversion. Any noise emerges as output least significant bit differences without signal distortion or phase delay errors, resulting from transmit buffer rate control and truncation. The single-channel ADEX transmitter includes six modules, listed as a least significant bit truncator, a compression filter, an entropy coder, forward error corrector, and a transmit buffer.

Incoming PCM continues through an LSB truncation stage where certain least significant bits become zero. The remainder continue on through an 8th-order, integer-based digital FIR compression filter for 16-bit deltas (differences). These now become word lengths of 3 to 46 bits, due to delta downshifting and entropy coding. Action continues for all 16-bit samples until accumulated bits constitute a block, which is "set aside" for format and error correction prior to transmit.

Because coding rates are variable, blocks have different samples, and more or less than normal blocks/second can saturate or fail to fill the transmit buffer, generating a rate control requirement. LSB truncating and downshifting supplies this need, and one bit truncated decreases the encoder data by about one bit/sample. This does reduce audio fidelity during signal peaks, and is the *only* noise or distortion contained in ADEX processing, according to Digideck.

Optimal or best parameters are determined by the first 32 samples as transmitter elements are adapted one per block, with parameters added as header data for receiver decoding. Such blocks are of fixed length in 17 packets of 64 bits, including a header packet with 48 bits for adaption and startup parameters plus 16 ECC bits, in addition to 16 data packets, each having 52 delta bits and 12 ECC bits. Right or left subframes consist of 8 to 16 blocks, depending on data rates. Two subframes constitute the 29.97 Hz frame, except that the 650 kbps direct system requires 10 blocks per subframe.

ADEX Receiver

As illustrated in Fig. 4-1 an ADEX receiver needs no LSB truncation or feedback truncation logic, therefore it has four inverse modules for decoding in addition to a muting and masking module that receives feed-forward from the ECC decoder. These consist of a receive buffer, ECC decoder, entropy decoder, and expansion filter. The masking unit covers over uncorrectable errors.

The receive buffer accepts the digital data stream, passing it on to the ECC and entropy decoders. The entropy code has a full range of 16-bit delta 1:1 conversions for positive value decoding of all difference values. The expansion filter has components with exact inverses to those of the compression filter, and so yields the same PCM samples on output. Therefore, no artifacts, no noise, and no distortion (except that possibly due to truncation) appear, and the receiver produces a replica of the original transmitter PCM input unless there are signal errors.

Errors

Analog audio has a stronger signal than video and is often heard even after video disappears. A compatible NTSC transmission could have multichannel sound in reserve should digital audio fail. Nonetheless, EDTV and/or HDTV audio proponents will have to design for worst-case conditions, with any drop-outs following and not preceding video fades. But until audio subchannels are firmly implanted in advanced television designs, possible directions are probably more supposition than clairvoyance with a few goblins mixed in.

Regardless, Digideck designers are looking at a bit error rate (BER) of at least 10^{-3}—well below that of picture fading that seems to take place around audio BERs of 10^{-5}. With such a goal, interpolation in unsuitable, and fixed-length data blocks, block codes, and some interleaving to deflect or overcome bursts. In the 650 kbps direct version, channel burst protection amounts to 167 bits, and in the augmented design, protection would be 121 bits. Engineers say this offers a raw bit error rate of 1×10^{-3}, which will approximate 10 block errors per minute, and a single frame error every 5 minutes, both of which require masking. Where there is a bad block or frame, masking is required to insert previous information "over the missing sample points." This actually means decoding one additional block.

For system startup, 9 parameters are required, including the final 8 samples of each frame that condition the 8th order digital expansion filter. The output receive buffer is next initialized by the transmit buffer's percentage of load, producing a calibrated delay between the I/Os. Following deinterleaving, entropy decoding, error correction, etc., the final 8 samples are loaded into the expansion filter, and the deltas are processed backwards (in time). Following subframes are then processed in forward time, and output samples are loaded, with actual output commencing when the receive buffer's output pointer reaches the S-sample end of the buffer, and when the corresponding video frame is able to display.

System Performance

Digideck claims very subtle degradation, no distortion, pumping, raspiness or grating on "even the most difficult material." Only at high levels of compression are there perceivable problems such as cymbal "hashing" or narrowing of sound spread. If original material is available, some differences could become apparent to the trained ear, but to the average listener, the effects are hardly noticeable.

The 15 KHz Augmentation System

This is the "improved" narrowband NTSC signal with additional detail in an augmentation channel, producing two 475-kilobit channels. The augmentation transmitter, illustrated in Fig. 4-2, generates a pair of outputs from the 44-kilobit digitally-converted audio. One is lowpassed and downsampled to 31.5 kbps, then ADEX processed around 6 bits per sample. The other signal becomes the difference between the original signal that's delayed and the initially downconverted signal. The reverse process, including upconversion, occurs in the receiver, with basic and augmentation signals added just before the digital-to-analog converter. Following D/A conversion the digital stereo is now ready for analog amplification and speaker outputs.

Digideck claims that 6-bits per sample sound quality will be better than existing commercial 15 kHz reproduction, that adding the two channels offers some 16 bits per sample, and that only a computer can find any least significant bit differences.

The augmentation receiver (Fig. 4-3) would need stereo pairs for decoding and the upconversion discussed. Decoding costs, it is said, will be relatively nominal compared to CD players with their $4 \times$ oversampling.

As for the future, a digital PCM system can provide four 15 kHz mono channels in the same spectrum as a 20 kHz pair, digital decoders for matrix signals offer superior performance, digital equalizers for receiver speakers, and eventually, entire rooms.

And one final suggestion would be to have a matrix-encoded system transmit foreground and background material separately rather than pre-mixed. One might then determine a personal mixture level consummate with his or her own hearing abilities.

The combination of digital video and digital sound does appear inviting, doesn't it? At this point, some of the more intricate details will be omitted, since you've probably had enough for now; but later we may add additional Digideck material. You may also want some material on BTSC-dbx multichannel

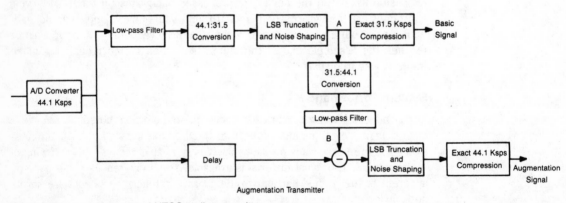

Augmentation Transmitter

Fig. 4-2. Improved narrowband NTSC audio transmitter (courtesy Digideck).

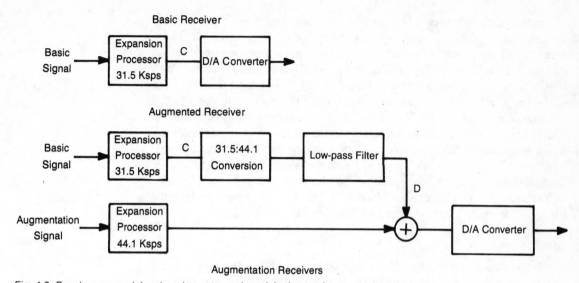

Fig. 4-3. Four inverse modules decode, mute, and mask in the receiver (courtesy Digideck).

stereo/SAP (second audio program) analog sound since this may well become a "fallback" proposition if or when super digital faults or fails. It will most assuredly be forthcoming after the Dolby description that follows.

THE DOLBY LABORATORIES PROPOSAL

The Dolby proposal incorporates three complementary system technologies including surround sound processing, digital audio sound coding, and digital data transmission. Four audio channels encompassing left, center, right and surround (L, C, R, S) channels would be matrix encoded using the Dolby 4:2:4 Motion Picture (MP) matrix for full decoding in the home. The two-channel matrix encoded information would be digitized using Dolby's adaptive delta modulation (ADM) coder and modulated for transmission using quadrature phase shift keying (QPSK). In the receiver the digital data would be recovered with a QPSK demodulator and converted to an analog signal by the Dolby ADM decoder. The original L, C, R, S audio signals would be recovered using a domestic Dolby Surround matrix decoder.

The Dolby ADM sound coding system introduced in 1984 was specifically designed for the American direct broadcast satellite market (DBS), which has yet to develop. ADM is now used, according to Dolby, for professional audio applications in satellite systems on several continents including North America and Australia. Audio quality is said to be high at low bit rates, decoder cost and complexity are low and the system is "tolerant" of bit errors in received data. A complete two-channel ADM decoder "can be built today for approximately $5.00—less in volume production."

In the ADM system encoder, audio is spectrally shaped and compressed in dynamic range before quantization by the system's core delta modulator. Three data streams are generated per channel: audio data, step size control data and emphasis control data. In the encoder, compressed analog audio is oversampled at 204.550 kHz (13× horizontal scan rate of 15,734.64 Hz) and quantized. Each of the two control data streams—one for volume range, and one for spectral shaping—are sampled at 7.876 kHz (half the line scan rate) and conveyed along with the coded audio for fully complementary restoration of dynamic range and frequency response in the decoder. At the operating data rates recommended by Dolby, the overall bit rate required for transmission of two audio channels would be 440.59 kilobits per second. Dolby also proposed providing transmission capacity for an additional 64 kbits/sec for auxiliary data carriage bringing the total transmitted data rate to some 512 kbits/sec.

QPSK

Dolby proposes quadrature phase shift keying (QPSK) for bit stream modulation on an additional digital sound carrier 4.85 MHz *above* the video carrier at a level 20 dB below peak unmodulated video. The QPSK signal is said to be rugged and easy to demodulate, but the additional sound carrier would fall 100 kHz outside the existing 6 MHz television channel slot and would intrude into the lower vestigial sideband of any adjacent channel. Possible complications include interference to FM radio channels 88.1 and 88.3 MHz operating in the same geographic area as a television station operating on channel 6. Similar interference potential into other assigned frequency bands exists for television channels 13 and 69. Other problems with the addition of the 4.85 MHz QPSK subcarrier relate to transmitters with notch diplexers that may need modification or additional tuned cavities to accommodate the QPSK carrier and to cable TV systems that modulate their FM sound carriers with AM timing pulses to key descramblers. Such systems are not compatible with demodulation of the QPSK signal using conventional intercarrier detection.

Recent testing of a similar proposal for use in Scandinavia, where the spacing between the aural subcarrier and the lower vestigial sideband of an upper adjacent channel video signal is identical to the situation in the U.S., suggested that the QPSK carrier could be added with minimal impact to upper adjacent channel video performance. In general, interference produced by the QPSK carrier into upper adjacent channel video or into existing FM radio channels at the low end of the band would be in the form of added noise. In particular, in older television receivers with LC vestigial sideband filters, some slight QPSK induced disturbance may appear in the form of noise in the picture; however, most current receivers "seem to be compatible" with the use of QPSK.

Quadrature phase shift keying is best described as pair bi-phase modulations set up in quadrature. As Fig. 4-4 shows, data enters a pair of low-pass filters at the operating data rate of 256 kb/s. Current bit values are combined

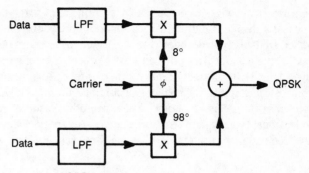

Fig. 4-4. Basic diagram of QPSK modulation (courtesy Dolby Signal Processing and Noise Systems).

into a single symbol which is representative of the two bit combination. The output symbol rate is 256 kb/sec with each symbol carrying two bits. In theory QPSK modulation yields 2 bits/Hz bandwidth efficiency. In practice however, some excess bandwidth is required to minimize inter symbol interference, minimize detection errors caused by timing jitter, and to allow for construction of practical Nyquist data filters.

MP Matrix

Dolby 4:2:4 Motion Picture (MP) matrix processing is used to encode the soundtracks of all current motion pictures released in Dolby Stereo. MP matrix processing constitutes encoding *four* primary sound sources (left, center, right and surround) into two optical soundtracks (Lt, Rt) on 35mm optical release prints. Adaptive dematrix processing is used to recover the four L, C, R and S signals during playback in Dolby Stereo equipped movie theaters. When Dolby Stereo films are transferred to video cassette and to laser disc the Lt/Rt matrix encoded soundtrack information is preserved. With the use of a Dolby Surround or Dolby Pro Logic Surround (full adaptive matrix processing) decoder the same spatial effects heard in the movie theater can be reproduced in the consumer's home.

MP Matrix processing is designed to be compatible with both conventional two channel stereo and monophonic playback. This is done by encoding right and left channels "without alteration" and placing center channel information in phase with both to produce a proper "phantom center" when no center speaker is used. Surround information is encoded out of phase as difference information between the Lt and Rt signals.

When a surround encoded signal is played back in conventional two channel stereo without benefit of the spatial decoder, a phantom center image is created for listeners equidistant from the two loudspeakers. Surround information is reproduced as an ambience effect. Do observe, however, that phase signs in the Lt and Rt channels are reversed. If a stereo signal is summed and reproduced in mono through a single loudspeaker, the out-of-phase information cancels and the surround effects vanish.

Available spatial decoders using the Dolby system are of two types: passive and active. Passive decoders extract L-R information, delay, and then process it with a modified Dolby B-type decoder (Fig. 4-5). Active decoders operate similarly but contain the additional adaptive directional enhancement circuitry that is used in playback for Dolby Stereo films in movie theaters. Adaptive matrix processing increases the "apparent" channel separation. Today's active consumer decoder, known as the Dolby Pro-Logic series, reproduces sound quality much like the professional type system used in cinemas. A center channel output is included to "restore proper lateral localization," of dialogue with a television picture.

Adaptive Delta Modulation (ADM)

Dolby claims that conventional pulse code modulation (PCM) needs 350 to 1,000 kbits/sec of data in each audio channel and require rather complex filters to achieve good performance. Under adverse error conditions, their engineers say the PCM systems require sophisticated and complex error correction schemes to maintain good performance. Therefore Dolby believes that delta modulation, an inherently simple system, has advantages. But linear delta modulation had insufficient dynamic range for high quality audio, and so a novel form of adaptive delta modulation (ADM) was developed.

One advantage of ADM is oversampling that allows a reduction of complexity in Nyquist filters. Another advantage is the possibility in an oversampled system of employing a technique called noise shaping. In a well-designed digital codec, quantization noise is distributed over its Nyquist bandwidth. In delta

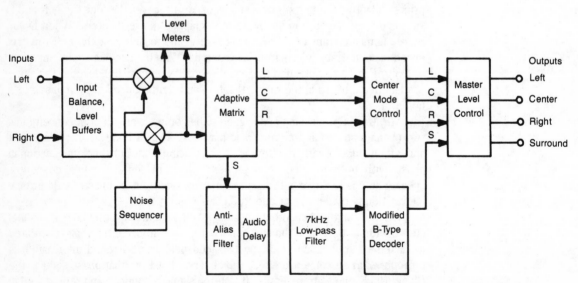

Fig. 4-5. L and R inputs are noise-reduced processed into four outputs for surround sound (courtesy Dolby Signal Processing and Noise Systems).

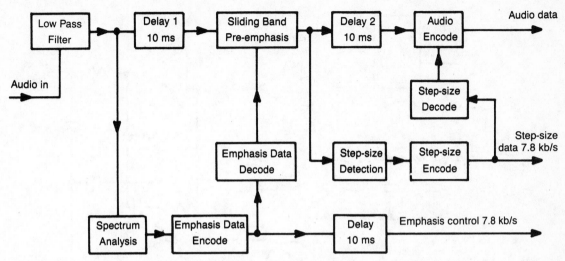

Fig. 4-6. A first class delta modulation encoder with low bit rate control (courtesy Dolby Signal Processing and Noise Systems).

modulation oversampled audio, most noise power is outside the audio frequency band and noise shaping can be used to redistribute quantizing errors to limit in-band noise and increase out-of-band noise proportionally. This particular ADM system chugs along at some $6 \times$ oversampling, and noise shaping is employed to decrease the noise spectrum at 6 dB/octave below about 15 kHz. Quantization noise at 1 kHz is reduced about 15 dB.

Reproduced errors in an ADM system tend to be less objectionable due to the fact that all bits are of equal weight. In PCM systems reproduced errors could be the result of an MSB "hit" and would sound objectionable.

As for quantization noise, when prime spectral components of the desired audio signal are below 500 Hz, high frequency pre-emphasis and de-emphasis help reduced quantization noise. Middle frequency signals between 1 kHz and 3 kHz can offer some low frequency noise masking, but the adaptive emphasis needs to rise to provide high frequency noise suppression and keep step size from becoming too great. Above 3 kHz, high level signals can be effective at masking high frequency noise, but noise at low frequencies needs to be suppressed to remain inaudible. Under these conditions a notch in the pre-emphasis curve will reduce the required step size and quantization noise. The complementary de-emphasis "selects" the desired signal components and "rejects" both low and high frequency noise.

The ADM decoder is said to be rather simple and is represented by the simplified block diagram in Fig. 4-7. Channel decoders are clock driven and receive emphasis, step-size, and audio data inputs. Audio data rates are "typically" around 204 kbits/sec, while step-size and emphasis control information are input at about 8 kbits/sec each. Step size and emphasis control information

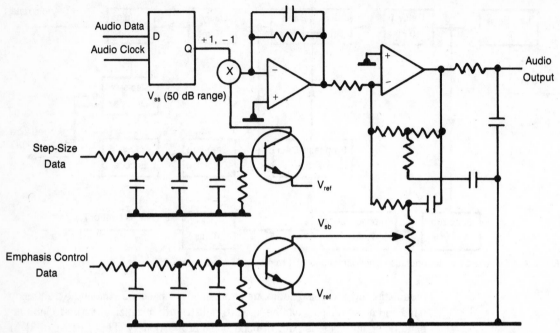

Fig. 4-7. Simplified diagram of the clock-driven and step-size decoder for delta modulation (courtesy Dolby Signal Processing and Noise Systems).

is decoded by low pass filtering to recover average signal values. Emphasis control governs the pole frequency variations in the de-emphasis network. When incoming audio data is multiplied by the decoded step-size control signal, large dynamic range audio is reproduced. De-emphasis then follows. Out-of-band components are attenuated by a simple low-pass filter and any necessary amplification takes place delivering music/voice sound to appropriate audible transducers. System hardware and integrated circuit decoders are said to be readily available according to Dolby.

At the beginning of 1989 the foregoing represents a fairly reasonable description of what these two digital-processing audio systems can and will do. If there's more to add later, including any definitive tests of positive mating with certain video counterparts, the information will appear at the end of the chapter, following our MTS explanation next.

MTS or BTSC-dbx

Specifically identified as broadcast television stereo/ SAP sound, this has been the prime TV broadcast version of stereo and second audio program material since FCC approval March 29, 1984. Today there are some 450 broadcast stations in the 50 states carrying such program material, and virtually all top-of-the-line television receivers are substantially equipped to decode and translate it into audible sound waves. You should see these same numbers—possibly better—in HDSTV by 1995. Sorry it can't be sooner, but the evolutionary process is both tedious and slow. The added "S" in HDSTV stands for sound, naturally. Also, BTSC represents the Broadcast Television Systems Committee, and *dbx* is the name of the company chosen to supply dynamic compression and expansion for noise reduction in the system. Keep in mind, however, BTSC-dbx is analog stereo rather than digital, therefore there's no similarity between the two; just reasonable compatibility, the changeover will probably continue well into the 21st century for both transmitters and receivers.

Also identified by the nontechnical world as simply MTS, or multichannel sound, this is a combination of FM and AM techniques, requiring additional carrier frequencies to trigger stereo or second audio programming SAP at either double or 5× the 15,734 Hz pilot carrier that's the same frequency as normal TV line scan rates. It's easy to generate and there's no interference between carrier and sweep because the pilot tone trigger is phase-locked to the transmitter's horizontal rate, and the tone, itself, is emitted within the FM 4.5 MHz passband.

Some 24 years prior to FCC approval, the National Stereo Radio Committee (NSRC) in 1960 began to study possibilities of AM, FM and stereo TV broadcasting. But the TV medium was not top priority because FM showed promise for best quality sound transmission. The same was largely true for AM stereo, which literally languished on the back burner until the late 1970s when the FCC hiccupped a couple of times with notices of proposed rule making. They finally approved the Magnavox system on April 9, 1980, but rescinded this ruling due to industry objections. Finally, on March 4, 1982, after a full-time review, the FCC finally decided to leave the entire issue to a "free marketplace"—a drastic decision that has taken over six years for "the market" to finally settle on Motorola's C-QUAM as the prime system for the U.S. and several other nations, but at the cost of possibly millions of dollars to everyone involved, as well as very slow AM stereo broadcast penetration. As of 1989, however, we are happy to report, C-QUAM is now firmly established in 12 foreign countries and 700 radio stations in the U.S. Canada decided and published its decision to base system approval on C-QUAM March 21, 1987—indicative of how "the world turns" when subject to a timid U.S. Federal bureaucracy imbued with the deficit-devised spirit of wholesale deregulation and an estimated FCC Washington lawyer-to-engineer ratio of 3:1. If you're interested in a vivid example of non-enforcement, just listen to Citizens Band Radio (CB) trucker talk, music, and profanity on the New Jersey turnpike any weekday of your selective choosing. Some of you may remember the days when call letter-numbers such as KAXP5047 meant you were complying with CB's voice-only law and freely identified yourself to prove it. Too bad the Judiciary thought $1 was too much to pay for an enforceable piece of paper then known as a license. Perhaps today's Supreme Court might hold a different view.

FCC Selection Process

Somewhat more resolute was the FCC's approach to multichannel sound, although naming a single system outright still wasn't yet tour de force nor de rigueur. But backed by the Electronic Industries Association (EIA) and an intensive series of tests, the FCC finally accepted the proposal of the Zenith Electronics Corp. somewhat indirectly by specifying that: TV broadcast stations may broadcast stereo sound with a subcarrier on the aural carrier, and the main channel having a stereo sum modulating signal and the subcarrier delivering a stereo difference signal . . . the subcarrier being the second harmonic of a pilot signal at the horizontal scanning frequency of 15,734 Hz $\pm$ 2 Hz, and double sideband amplitude modulated with suppressed carrier and capable of "accepting a stereophonic difference encoded signal over a range of 50-15,000 Hz." Transmissions of a second audio program (SAP) were also authorized, with subcarrier frequency of $5\times$ the horizontal line rate and peak deviation of $\pm$ 10 kHz. Professional low frequency carriers were also approved for data up to and including approximately 100 kHz, all multiples of the pilot signal above the second harmonic with the exception of SAP at 5_H—the highest being 6.5_H, or 102.271 kHz.

In its award, the FCC specified that any system using the 15,734 Hz subcarrier must broadcast and receive the BTSC-dbx method of processing stereo and SAP. That, apparently, has been the Zenith system lock. Using this as a somewhat tenuous example, it's possible the FCC might well follow a similar route in approving an NTSC compatible method without naming an absolute winner. There are undoubtedly enough lawyers at 1919 and 2025 M Sts. in Washington to devise another convenient circumlocution that will probably work; we'll have to wait and see.

In the meantime, visions of sugarplums for ATTC testers could turn into sour grapes as video systems are forwarded for testing in the middle (or later) of 1990. Most of them are leaving "slots" for digital audio here and there, attempting to place some of the burden on either Digideck and/or Dolby. ATTC attempted to require the submission of complete systems, but the video proponents may have different ideas. These and other outstanding questions are going to need more than a little resolving/revolving before substantial results with definitive answers are possible. Audio companion systems, however, *are* required.

Obviously the FCC will want numbers and advantages/disadvantages of the several surviving systems, and digital audio placement and performance is decidedly one of them. Beautiful pictures and no sound won't elicit overwhelming interest from John or Jenny listener who's just shelled out $2000 or more for fancy HDTV with no S in the middle. True, the accepted interest ratio of 70:30 for video/sound remains the quasi-accepted standard, but both are required for general enjoyment, especially when big bucks are laid on the table.

BTSC-DBX, A WORKING EXAMPLE

We can cite system ideas and objectives the day long, but there's no substitute for a few diagrams and specific analysis of a representative multichannel sound hookup that's now fully operational. Here we'll choose a Philips' Magnavox as a worthy example: specifically, a PTV100 projection receiver with a 37-in. diagonal screen having an ultra-wide angle viewing screen of 180° expanse with considerable brightness and a vertical viewing angle of 28°.

Were this a pure television book, we'd do the entire theory of operation (with the usual assistance from Magnavox), but it isn't, so only the BTSC-dbx portion rates analysis. But you should know the receiver has 11 panels and 7 labeled modules—all pluggable/unpluggable; so it's quite home serviceable, which is a singular asset for these large, complex television receivers, most of which are not!

The stereo/SAP decoder is one of the seven modules. It contains amplifiers solid state switches, comparators, expanders, buffers, and a pilot signal detector—a large Siemens surface wave acoustical filter (SAW) separating video and audio i-fs into their respective 45.75 MHz and 41.25 MHz frequencies, although the sound detector uses both video and audio beat differences to resolve the 4.5 MHz intercarrier from which all FM audio derives. For this example, we'll use one of Maggie's excellent block diagrams (Fig. 4-8) that details much of the decoding action, but not the final amplifiers whose inputs derive from the I/O panel that the decoder module feeds.

Once the carrier frequency has been reduced and 100 kHz wideband composite audio is available from the sound detector, such signals are routed to the IC1 stereo decoder where monophonic is buffered by an L + R output, processed through a 0-13.5 kHz low-pass filter and on to amplifiers and outputs.

If the signal is stereo, however, such signals are buffer-amplified, the pilot detector energized, a Schmitt trigger flips the stereo-mono switch, and L-R information flows into the stereo decoder. At the same time the AFPC phase detector also receives the pilot detector signal, pilot and local oscillator signals are compared, and any necessary frequency/phase correction generates true tracking for the 189 kHz oscillator and control. There is also an R35 frequency adjust available for manual tuning, in addition to a divide-by-12 circuit that delivers LO oscillations to the L-R stereo decoder. L + R and L-R outputs are now leveled via their respective potentiometers, with L + R continuing through resistive dividers R91, and R92 to the negative inputs of the two operational amplifiers in IC4, while L-R must transmit through the dbx expansion and quieting portions of IC5 and IC6 before rejoining L + R at the positive inputs of IC4 for IC6 switching into the L/O panel. In IC1, the decoder, there is also an output from the stereo mono switch to the tuner interface and LED panels that energize the stereo light emitting diode lamp on the front of the receiver. Before tackling the dbx quieting portion, however, let's backtrack a little and describe

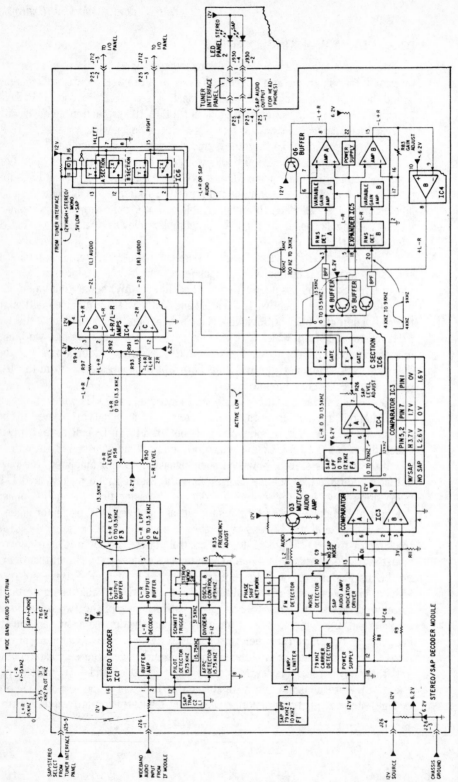

Fig. 4-8. Magnavox/Philips MTS decoder signal flow block diagram of the BTSC-dbx system in a TV consumer product (courtesy Philips Consumer Electronics).

the second audio SAP operation since both L-R and SAP must pass through this final circuit before continuing on to the I/O outputs.

Should SAP be received instead of stereo or mono, the L1, C2 SAP trap shunts the 78.67 kHz carrier and its modulation to ground prior to entry into IC1, but has no effect on SAP's direct path to its 79 kHz bandpass filter below that has a passband of 10 kHz. Being FM and not AM sideband information, the SAP input is demodulated by an FM detector within the module. But should the signal become noisy, it is coupled back into a noise detector, transistor Q3 is biased off, and noise is prevented from reaching the SAP 0-12 kHz low-pass filter and the earphone output at P25.

Once noise passes, audio information continues on through amplifier IC4, the X-Y switch gates for the amplifiers in IC4 above, and on to the I/O panel for eventual final amplification and speaker outputs. On the way, however, sound signals must also pass through the dbx companding circuits in IC5, whether stereo or SAP—a condition not true for any ''professional'' lower frequency data information.

The dbx Operation

As illustrated by IC5, a pair of emitter follower buffers supply signals to the A and B rms detectors, their variable gain amplifiers, (Fig. 4-9) and final A-B amplification, with B feedback through IC4 into the expander through pin 5. Note that inputs into rms A range between 3 kHz and 100 Hz, and inputs into rms B are listed from 4 kHz to 9 kHz; therefore, the pair of detectors, since they handle completely different frequencies. These two complementary decoders allow dbx to deal with many audio problems such as hiss, buzz, whistles, and hum. The encoder restores spectral and amplitude dynamics to the transmitted signal by amplitude expansion and static de-emphasis of 1:2 at low and mid frequencies, increasing to 3:1 at high frequencies. Approximately 8 dB is allowed as ''headroom'' for static information and clipping occurs only for small-energy transients with virtually no signal attenuation.

Some receiver manufacturers, however, deemed this insufficient when MTS was first introduced and added a National Semiconductor dynamic noise reduction circuit for further quieting, identified as the LM1894. This offered bandwidth control paralleling the ear's sensitivity to noise accompanied by music via a low-pass filter and an integrating OpAmp.

Sound Thoughts

Regardless of the secondary heading above and its somewhat questionable implications, you now have much of what is to be involved in the HDSTV of what's to come. Industry has definitely decided that good video deserves equal-valued audio, and the only real way to do that is with 1s and 0s. Also, that the new system(s) have to be compatible with BTSC-dbx; and that's why we detailed the structures of both digital and analog audio operations.

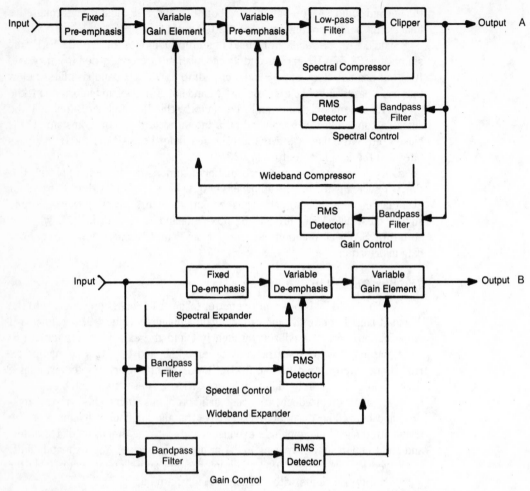

Fig. 4-9. The combined compressor and expander from dbx illustrated in block diagram form (courtesy dbx, Inc.).

Frankly, the sound quality available in discrete and multiplexed satellite stereo has been more than satisfactory up till now, when applied to a set of woofer-tweeter speakers with good crossover networks and reasonable baffles. The CD disk has also been excellent, and digital audio in 8-millimeter VTRs and camcorders should be remembered and used for its intended purpose. Should forthcoming high definition systems surpass those only recently invented, more power to them. But, in the meantime, we should try and enjoy what's here and not necessarily wait another several years.

We might add that Hughes Communications (and possibly others) have already begun HDTV satellite testing with several system proponents over either their Galaxy or Western Union orbiters acquired in the latter half of 1988.

According to Hughes, they are now America's leading distributor of cable and TV programming, having 144 C-band Galaxy and Westar transponders at their disposal. Philips and Hughes will test first, with others to follow. A few million for operating expenses apparently helps. At the moment that's about the sound track of the HDTV story. Yes, it's coming, and certainly the results should be remarkable—for a price.

5
IDTV: The Forerunner of EDTV and HDTV

THIS CHAPTER IS ENTIRELY DEVOTED TO THE LATEST IN *IMPROVED DEFINITION* television receivers and camcorders, they are actually the predecessors of their much more glamorous relatives. So while we develop, test, modify, and wait for the super new sets to come on line, IDTV electronics can certainly give the critical a considerable foretaste and understandable introduction to that ultimate technology just months away. And before 1990 and 1991 join the grim reaper, you'll certainly read about and possibly see both satellite HDTV and some experimental terrestrial broadcasts as well. Even if the final system(s) isn't or aren't yet chosen, there's no reason to assume that the "perfect picture" search is dormant; obviously it isn't. And a great deal of activity will be under-taken prior to the FCC's ragged or carefully considered blessing.

Certainly the interval between IDTV and HDTV is sure to produce a scramble between U.S., European and Japanese distributors and manufacturers to introduce more advanced designs with exceptionally attractive features. Field and frame store technology, double horizontal scan rates, artifact suppression, digital video and audio processing, vastly improved on-screen control functions, virtually perfect luma/chroma convergence, 40-inch (perhaps even larger) cath-ode ray tubes with astonishing resolution, almost amazing recorder and TV sound reproduction, much improved video in VCRs and especially camcorders in both VHS and Beta formats. And probably before the end of 1989, you'll see this excellent video advance translated into the 8-millimeter format that has already proven a natural for lightweight camera and recorder systems the enter-tainment world over.

Would you be surprised to know that almost all these special engineering advances already exist? If there's any question, do read on, for that's what this chapter is all about—an introduction to very solid consumer IDTV products available right now! For if you're satisfied with 4 MHz video and 15 kHz stereo sound just as it's broadcast, then a fat checkbook and a trip to the store is all that's required. But don't forget the TV antenna and coaxial cabling that goes along with it; and don't expect poor CATV systems to suddenly deliver superior images and sound—they just don't cut the mustard. Otherwise, you'll see and hear exactly what's transmitted, be it good, bad, or evil. The choice is entirely yours!

IDTV BY PHILIPS

Designed and engineered by Sai Naimpally, T. Darby, L. Johnson, L. Phillips, and J. Vantrease of Philips Consumer Electronics in Knoxville, Tenn., this system is a truly improved definition version of standard NTSC with features that will certainly carry over into HDTV in one form or another. Initial receiver shipments occurred in the fall of 1988 in 24-inch table models and 31-inch consoles.

This new consumer-product receiver offers non-interlaced scan conversion, field comb filtering, video noise reduction, still pictures, store and recall, 3- and 9-PIP strobe and preview, all with a single field memory. Visible picture improvements include line crawl and line structure elimination, less line flicker, additional sharpness and resolution, reduction in interference and lessening of hanging dots and dot crawl. Strobing permits the capture of successive pictures while preview displays up to nine different channels. The single field memory and recursive filter are said to "clean up" weak signals, reduce noise up to 10 dB, coupled with motion compensation. Digital memory claims 7-bit resolution at a 9-MHz sampling rate, and hardware further includes double-sided printed circuit boards, surface-mounted integrated circuits, and custom digital ICs, all developed in Knoxville. These new receivers also have an S-VHS extended definition input, audio/video input and output jacks, a variable audio output and 3-position RF switcher, in addition to a multiproduct controller that can "learn" the routines of most VCRs, cable converters, and audio plus auxiliary codes not preprogrammed.

As you can readily see in the block diagram (Fig. 5-1) with its PIP and video inputs, the system is somewhat complex and certainly full-featured, even among the separate stages of IDTV processing.

Composite picture information proceeds into the main video port, is digitally encoded with a 7-bit A/D converter, digitally decoded with the help of a 13.5 MHz line-locked clock and routed to picture-in-picture in 4:1:1 format with luminance clocked at 13.5 MHz and color difference signals at 3.375 MHz. With a 12-bit decoder output, 7 bits carry luminance and the R-Y and B-Y information occupy 4 bits of 7. All this is converted to an 8-bit, 20.25 MHz bus by the PIP read gate, which later becomes a 12-bit bus again in the non-interlace scan circuit.

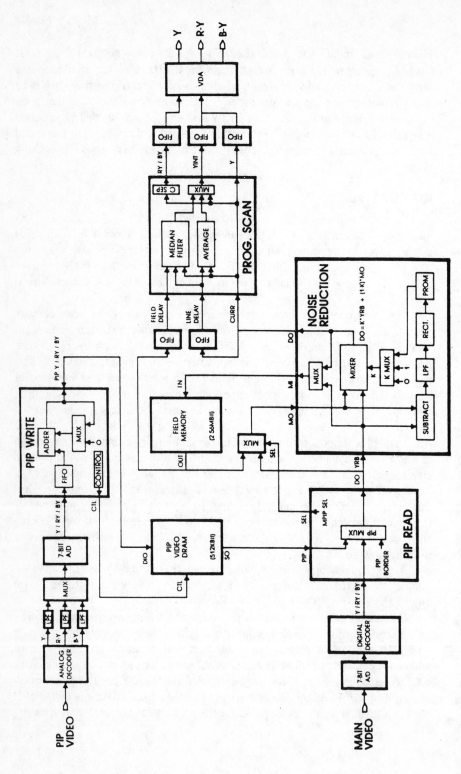

Fig. 5-1. A block diagram of the new Philips IDTV system available in both consoles and table models (courtesy Philips Consumer Electronics).

114

At the same time, PIP composite video enters the IDTV processor at the top left, is analog decoded into Y and R-Y, B-Y signals, low-pass filtered, multiplexed, and again digitally encoded by another 7-bit A/D converter for the PIP write circuit. Data then passes through a first-in, first-out (FIFO) buffer memory that, apparently, is a serial device feeding the adder above. When PIP is desired, it is selected by a digital multiplexer, including control and a digital random access memory (DRAM), bypassing main video.

With the reception of processed video from the main video input, the noise reduction and field comb filter eliminates crosstalk and adds a 10 dB noise attenuation. These are aided by a simple recursive filter and a motion detector for filter depth and serial memory of one field length for delay. Filter K-factor depth can be selected between $1/8$ and 1, with the mixer being responsible for recursive filtering within the noise reduction portion. The filter consists of a subtractor, a filter and a multiplier. The field memory subsystem is next, with inputs from the noise reduction MUX and outputs to a second MUX that inputs the noise reduction mixer. At a 7-bit data input from the noise reduction/field comb filter at 20.25 MHz, the information is stored a line at a time following set delay, data is again outputted at a 20.25 MHz clock rate, and when entering the stored picture mode, additional inputs are rejected and the memory will refresh data continuously until otherwise commanded. During the active control signal, the field memory accepts and transmits data, with outputs leading inputs by 35 clock cycles. An offset to the DRAM memory address appears every 294 command cycles plus multiplexer delays make the output leads possible.

The progressive scan arrangement follows two additional FIFOs whose inputs originate from the noise reduction IC and the field memory. These inputs are identified as field/line delays and direct input luminance, whose data rate is doubled by the Y FIFO. Y information also is routed to the MUX and C separator sections, as well as the averaging section and the median filter. Line data remains delayed by 1H, and field data is FIFO-buffered to compensate for alternate field one-line offset. Line, field, and Curr. (Y) information is then averaged, filtered, multiplexed and color-separated for RY/BY and Y int, the latter having a doubled data rate. Color difference signals above are used for both standard line and line repeat. Three FIFOs then deliver their outputs to the video D/A converter along with Curr. (Y) luminance. Analog outputs, as you can see, are color difference signals R-Y and B-Y, plus the usual luminance.

IDTV BY ZENITH

This is the second version of an improved definition receiver by Zenith, like the first one, it's all digital from the video detector to D/A outputs for video and sound. Improvements, however, are both considerable and numerous, and this No. 2 edition produces as clean an NTSC picture as you virtually can desire, and also contains vastly improved sound, even without super Bose (expanded) ''theatre'' audio delivery.

It's all based on ITT Intermetall's extensive developments in Freiburg, West Germany and the Digit 2000 VLSI digital TV system designs. In the past year alone ITT Semiconductor has announced a new progressive scan processor, video A/C converter, an improved multiclock generator, special deflection processors for text and D2-MAC (multiplexed analog components), an updated NTSC comb filter and video processor, and special high-speed A/D/A converters of the flash type, as well as a new teletext processor for Level 1 teletext capable of serving both Germany's PAL (phase alternating line) and our own NTSC. Quite a bundle of achievements that are steadily moving into several digital TV systems both here and abroad.

Zenith, fortunately, has been a subscriber to ITT's efforts and is now able to install many of these new ICs in an excellent U.S. receiver that appears to have virtually no video artifacts, very good digital audio, and resolution/definition even from standard broadcasts that has created more than considerable attention. Coupled with its new HDTV system and improved JVC-manufactured camcorders, the Zenith of old may well return to its stellar spot in video entertainment markets. We're betting on the last of the U.S.-owned manufacturers "to do it again" as the year 2000 draws ever closer.

New VLSI circuits directly referenced by Zenith include four, which are identified in technical training manual TP41 as the MCU-2600 clock generator, VAD-2150 video A/D converter, the CVPU-2230 video processor, DPU-2543 deflection processor, PIP-2250 picture-in-picture processor, a 64K × 4 dynamic RAM, and a number of supporting, smaller logic ICs. For some strange reason, the TPU 2732 Teletext processor is not listed, although available and installed in the new models.

Receiver Programming

Simply enumerating and describing the ICs alone tells only part of the story. Functional controls and the results they produce contribute the other half of what you will shortly realize is one remarkable receiver, well worthy of U.S. design and production. And even though the digital portion consists mainly of ITT's VLSI circuits, analog, power supplies, cathode ray tubes, and certainly power driven developments are primarily Zenith products, in addition to microprocessor controls that not only deliver menu audio/video/antenna call-up settings, but also Teletext and even triple picture-in-picture (PIP) displays. Furthermore, a "jackpack" (baseband audio video I/Os) on the back accepts demodulated sound and pictures from any normal consumer source with the usual 75 ohms and 600 ohms for related impedances. Once you hear even multichannel digital sound and see extremely clean, artifact-free pictures, you'll begin to realize what's in store when double-bandwidth HDTV and dual or quadruple digital sound arrives. In the meantime if you're anxious, try some good satellite transponder reception for both video and discrete or matrix stereo. Even that, combined with 4/12 RF GHz delivery from 22.3 kilomiles above the equator, is a substantial foretaste of what's to come. For now, enjoy full 4 MHz video and passably good sound from the TV stations.

Probably the first unique feature worth discussing is the *menu on-screen display* and execution. Microprocessor controlled, there are actually three different sets of menus, two of which are customer-available, and a third is reserved for factory and field technician setups. The first two will be discussed, along with auto demonstration, while the factory/field adjustments are withheld for strictly technical reasons and to prevent unauthorized tampering with basic memory loading. Regular programming may take place on either the receiver's 9-keyboard touch units or its infrared remote control that also programs both VCRs and Teletext, as well as PIP, if this function is included on specific models.

Factory programming generally places most microprocessor settings at mid-range and therefore any conventional changes must be derived directly from the main menu. This menu appears as indicated in Fig. 5-2. As you see, this includes the various inputs (sources), VCR commands, time sets, picture, audio, sharpness, color adjustments, and a (comb) video filter that you certainly want activated for best color and luminance separation, plus maximum luma bandwidth and constant tint and chroma amplitude control. Most of this is self-explanatory except items 23 and 24 which, translated, refer to surround sound and projection (RGB) alignment setup. Menu numbers 15 and 16 are used for

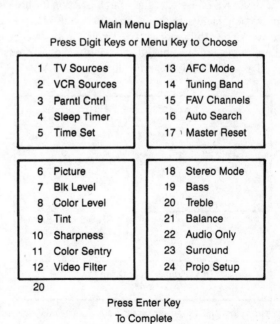

Main Menu Display

Press Digit Keys or Menu Key to Choose

1	TV Sources	13	AFC Mode
2	VCR Sources	14	Tuning Band
3	Parntl Cntrl	15	FAV Channels
4	Sleep Timer	16	Auto Search
5	Time Set	17	Master Reset

6	Picture	18	Stereo Mode
7	Blk Level	19	Bass
8	Color Level	20	Treble
9	Tint	21	Balance
10	Sharpness	22	Audio Only
11	Color Sentry	23	Surround
12	Video Filter	24	Projo Setup

20

Press Enter Key

To Complete

Fig. 5-2. Receiver tuning may be preset or selective at customer preference (courtesy Zenith Electronics Corp.).

117

programming favorite channels and automatic search where the latter will automatically find, program, and select all available channels in your area with reasonable signal strengths. If you wish to either delete some of these or add several that are a little weak, then Menu No. 15 will allow further specific selections. An on-screen selector bar under FAV channels will indicate whether it is saved or skipped following Enter and the Adjust Up/Down dual action button.

You may also set local time, activate a go-to-sleep timer, and program VCR sources in addition to parental control. This allows one channel to be "blanked" for a 12-hr period, that can only be activated during that time by your own designated 4-digit code. Of course if it's forgotten, you'll have a half-day's wait before full channel scan returns to normal. A detailed operating guide and warranty provisions should supply the remainder of any and all these Menu instructions.

In Fig. 5-3, however, we would like to briefly review the various options, many of you may not yet own one of these receivers. You can select CATV, standard outside TV antenna, loop through for a decoder, two VCR audio/video connections, a VCR Y/C input, an RGB S-VHS input, and the regular broadcast band. It's also worthwhile knowing that the automatic frequency control (AFT/AFC) has two modes: one fixed and the other search. In fixed, the variation from center channel frequency usually isn't that wide for either CATV or broadcast. But where cable systems have harmonically-related (HRC) or incremental coherent carrier (ICC) RF emissions, the search position for broadband channel acquisition is extremely handy. In this connection, you'll have to tell the tuner whether you're working with broadcast, CATV, HRC, or ICC by engaging No. 14 in the main menu display. And baseband video/audio also must be selected separately as well as S-VHS, RGB, and/or luma/chroma YC.

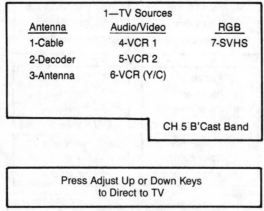

Fig. 5-3. Since this is a monitor/TV receiver, inputs from VCRs, CATV, Teletext, and RGB are all possible (courtesy Zenith Electronics Corp.).

In short, the microprocessor has to be correctly programmed and told what to do if you want satisfactory results. We don't mean to imply that complex digital receivers are extraordinarily difficult, but they aren't California flower children either. Once past the initial I/O plunge, you'll appreciate the enormous flexibility of this highly responsive receiver. On the receiver's back you'll find internal/external speaker selection, plus stereo and video in/out loop, and the CATV/ant. options (Fig. 5-4).

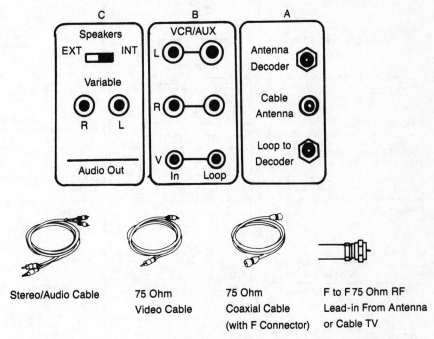

Fig. 5-4. Outputs are selective as are internal/external speakers, cable and loop through (courtesy Zenith Electronics Corp.).

Standard programming may take place at the receiver or, better yet via the remote control since it is equipped to offer all programming whether recorder, teletext, of TV/CATV. In Fig. 5-5, Zenith's drawings of both TV and teletext remotes are included, along with directed explanations. And rather than enter into a long-winded discussion of the various steps, we suggest you read whatever is of interest and ignore the rest. These explanations and directions are very adequate and to the point. Freeze frame and PIP are also included, as well as flashback between the last two channels tuned. What's more important, however, is how the system, itself, works, and the remarkable circuits that make it go. So this topic is covered next.

The Remote Control (TV Operation Only)

This page outlines operating your TV using the remote control. For instructions on operating your Zenith VCR and Teletext see Section III; "Using Your Remote Control."

Set the VCR/TEXT/TV switch on your remote control to the TV position.

For optimum performance, point the Remote Control toward the TV.

SPACE COMMAND 9500 TV FUNCTIONS
(MODE switch in TV position)

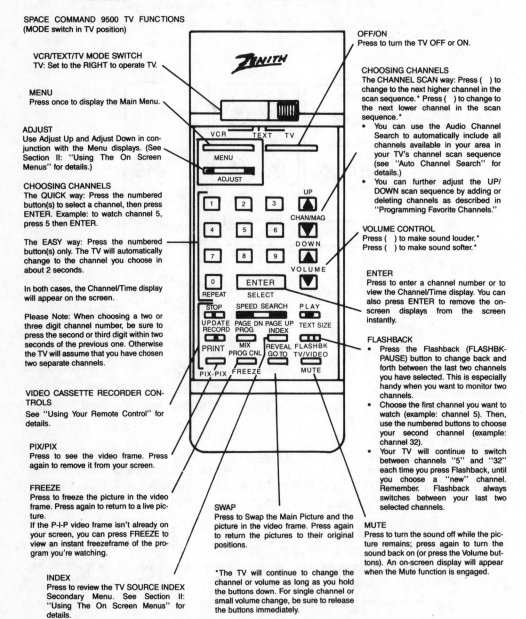

VCR/TEXT/TV MODE SWITCH
TV: Set to the RIGHT to operate TV.

MENU
Press once to display the Main Menu.

ADJUST
Use Adjust Up and Adjust Down in conjunction with the Menu displays. (See Section II: "Using The On Screen Menus" for details.)

CHOOSING CHANNELS
The QUICK way: Press the numbered button(s) to select a channel, then press ENTER. Example: to watch channel 5, press 5 then ENTER.

The EASY way: Press the numbered button(s) only. The TV will automatically change to the channel you choose in about 2 seconds.

In both cases, the Channel/Time display will appear on the screen.

Please Note: When choosing a two or three digit channel number, be sure to press the second or third digit within two seconds of the previous one. Otherwise the TV will assume that you have chosen two separate channels.

VIDEO CASSETTE RECORDER CONTROLS

See "Using Your Remote Control" for details.

PIX/PIX
Press to see the video frame. Press again to remove it from your screen.

FREEZE
Press to freeze the picture in the video frame. Press again to return to a live picture.
If the P-I-P video frame isn't already on your screen, you can press FREEZE to view an instant freezeframe of the program you're watching.

INDEX
Press to review the TV SOURCE INDEX Secondary Menu. See Section II: "Using The On Screen Menus" for details.

SWAP
Press to Swap the Main Picture and the picture in the video frame. Press again to return the pictures to their original positions.

*The TV will continue to change the channel or volume as long as you hold the buttons down. For single channel or small volume change, be sure to release the buttons immediately.

OFF/ON
Press to turn the TV OFF or ON.

CHOOSING CHANNELS
The CHANNEL SCAN way: Press ( ) to change to the next higher channel in the scan sequence.* Press ( ) to change to the next lower channel in the scan sequence.*
• You can use the Audio Channel Search to automatically include all channels available in your area in your TV's channel scan sequence (see "Auto Channel Search" for details.)
• You can further adjust the UP/DOWN scan sequence by adding or deleting channels as described in "Programming Favorite Channels."

VOLUME CONTROL
Press ( ) to make sound louder.*
Press ( ) to make sound softer.*

ENTER
Press to enter a channel number or to view the Channel/Time display. You can also press ENTER to remove the on-screen displays from the screen instantly.

FLASHBACK
• Press the Flashback (FLASHBK-PAUSE) button to change back and forth between the last two channels you have selected. This is especially handy when you want to monitor two channels.
• Choose the first channel you want to watch (example: channel 5). Then, use the numbered buttons to choose your second channel (example: channel 32).
• Your TV will continue to switch between channels "5" and "32" each time you press Flashback, until you choose a "new" channel. Remember, Flashback always switches between your last two selected channels.

MUTE
Press to turn the sound off while the picture remains; press again to turn the sound back on (or press the Volume buttons). An on-screen display will appear when the Mute function is engaged.

Fig. 5-5. Space Command remote control for TV, VCR, and Teletext (courtesy Zenith Electronics Corp.).

120

SPACE COMMAND 9500 TELETEXT FUNCTIONS
(MODE switch in TEXT position)

PAGE SELECTION
Press three numbered buttons to select a specific page. For example, to view page 150, press 1, then 5, then 0. After a few seconds, the page will appear.

UPDATE
Press to find out when a page has been updated. Press UPDATE only after the page you want has appeared. The TV picture will return and the page number will appear briefly in the upper left corner. When the page is updated, the top line of it will appear across the top of the screen. Press ENTER or REVEAL to see the updated page.

PRINT
Use this button if you have a Zenith Printer (option accessory) and you wish to print pages of Teletext. The page that appears on the screen is the one that will be printed.
    When PRINT is pressed, the picture will revert to the regular TV picture and one of the following displays will appear:

Indicates printing in progress.

Indicates printing is complete.

After printing is done, return to the Teletext display by pressing any text key.

An error message will appear when the Printer is not connected or is turned off.

TV/VCR/TEXT MODE SWITCH
TEXT: Set to MIDDLE position to operate TELETEXT.

OFF/ON
Once you have selected a channel that broadcasts Teletext, put the Mode switch in the Text position and press OFF/ON to initiate the Teletext display. While watching Teletext, press OFF/ON to temporarily switch to the regular TV picture. Press again to return to Teletext.

CH/MAG (CHANNEL/MAGAZINE)
Press to increase or decrease page numbers in 100 page increments. Once selected, the page will appear after a few seconds.

PAGE DOWN
Press to decrease the page number, page by page.

PAGE UP
Press to increase the page number page by page.

TEXT SIZE
Press to increase text size. Press once to view the top half (12 lines) of the screen; press again to view bottom half. Press a third time to return to the normal display.

REVEAL
Press to reveal an updated page or ''hidden'' answers.

MIX
Press once to view the Teletext display superimposed over the TV picture. Press again to return to normal Teletext display.

A Block Diagram

A block diagram (Fig. 5-6) pretty well outlines the system in signal flow dimensions, including the digital functions that will be discussed in some detail following a brief explanation here. As Zenith aptly remarks, ''cross a TV with a computer and you get digital TV.'' Essentially that's the sum and substance of what this is all about. You see the abbreviations MCU, CVPU, VCU, DPU, EAROM, RAM, Keyboard Microprocessor, and CCU. In order, they represent clock generator, NTSC comb filter and video processor, video codec (ADA conversion), deflection processor, electrically addressable read only memory, and random access memory. Analog functions are the tuner, audio/video i-fs, the audio, vertical/horizontal outputs, and the cathode ray tube. Below the i-fs is

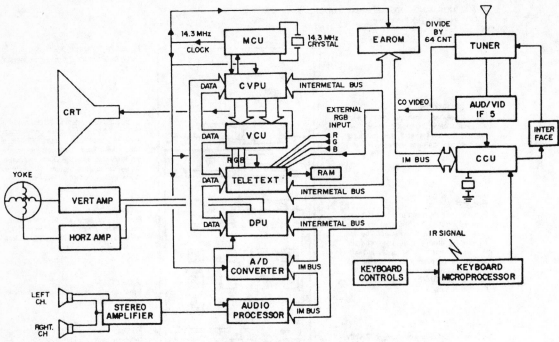

Fig. 5-6. System block diagram of the entire receiver, with digital emphasized (courtesy Zenith Electronics Corp.).

another digital IC, the central control unit (CCU) and picture-in-picture PIP that's not illustrated. With the exception of the CCU, we can fairly accurately identify these very large scale integrated circuits as belonging to the ITT Intermetall group out of West Germany. The microprocessor control is unfamiliar, as is the microprocessor. Discussions, therefore, will involve the Digital Main Board since that's where all conversions and digital processing occurs anyway. All else is strictly controls.

Composite video/audio enters from the tuner and i-fs via either direct inputs or through RGB connections at the Jackpack of the receiver. A system clock generator now takes over so that all digital operations will be synchronized, video and audio are digitized through A/D converters, the keyboard microprocessor selects an input for teletext, video, or CATV, the deflection processor control vertical/horizontal sync, and picture-in-picture circuits offer extra images in one or three frames, and then audio and video are D/A converted for final analog pictures and sound.

Analyzing the ICs

The next logical progression is a short analysis of these identified ICs, including as much of Zenith's main digital block diagram 9-700 as page space in the book will allow. This affords a representative idea of how these ICs interact with one another in a proven digital system that's already market-available and highly recommended. Each will be represented by both Zenith's and ITT's designations. Figure 5-7 (in two sections) illustrates both the logic and their connections.

The MCU Clock Generator IC identifies simply as IC 2301 with no special ITT number. Apparently, however, its clock frequency is 4 × 3.579545 MHz, the color subcarrier, has at least one voltage controlled oscillator, two phase-locked loop circuits, filtering, and an output of 14.318 180 MHz with tight crystal control. It's second cousin, ITT's MCU 2632 also has a 12-bit shift register, a D/A converter, a frequency doubler, and three voltage controlled oscillators for PAL, SECAM, and our own NTSC. The Zenith version is obviously not as extensive. It isn't required to clock for Germany's PAL or France's SECAM. For a possible "universal" TV, however, the MCU 2632 is readily available.

The CVPU 2233 is Zenith's IC 2202 NTSC comb filter video processing unit digitized by the VCU 2133 video codec. It's an N-channel MOS that includes a code converter, an NTSC comb filter, chroma bandpass filter, luminance filter with peaking, a contrast multiplier, all color processing, an IM bus, and CRT current levels. Digitized video is delivered in parallel Gray code by the VCU 2133 and converted to binary for the comb filter offering maximum separation of chroma and luminance. The luminance portion will accept either 4 MHz from broadcast signals or up to 7 MHz from baseband video, along with special peaking followed by a contrast multiplier that is controlled by microprocessor CCU (not otherwise identified). The CCU also permits chroma adjustments, while a PLL circuit maintains chroma phase for constant hue. Finally, Y, R-Y, and B-Y are all D/A converted, matrixed, and delivered as RGB outputs to IC 2201.

VCU 2133 serves as ITT's designator for Zenith's IC 2201. Identified as the video codec, this is a high-speed encoder/decoder for ADA video signal conversion. Another 40-pin VLSI, it contains two video amplifiers, one A/D converter, a noise inverter, one D/A luminance converter, two D/A color difference converters, an RGB matrix, output amplifiers, and programmable beam current limiting.

It converts analog composite video into digital information. RGB color signals are derived from color difference reds and blues when D/A converted and both luminance (in analog form) and color intelligence are matrixed for RGB outputs. Beam current and peak beam current limiting takes place in a sensing circuit from the RGB amplifiers which reduces D/A reference voltages for luminance and chroma, thereby controlling picture tube drives, especially contrast and brightness.

The DPU 2543, or IC 2701 is next. This IC is programmed for standard horizontal scan, Teletext and D2-MAC receivers. Within you'll find video clamping, horizontal and vertical sync separation, synchronization, and deflection, sawtooth ramp generation, E/W CRT sweep correction, text display circuits, and D2-MAC capability. Programmable characteristics include TV standard

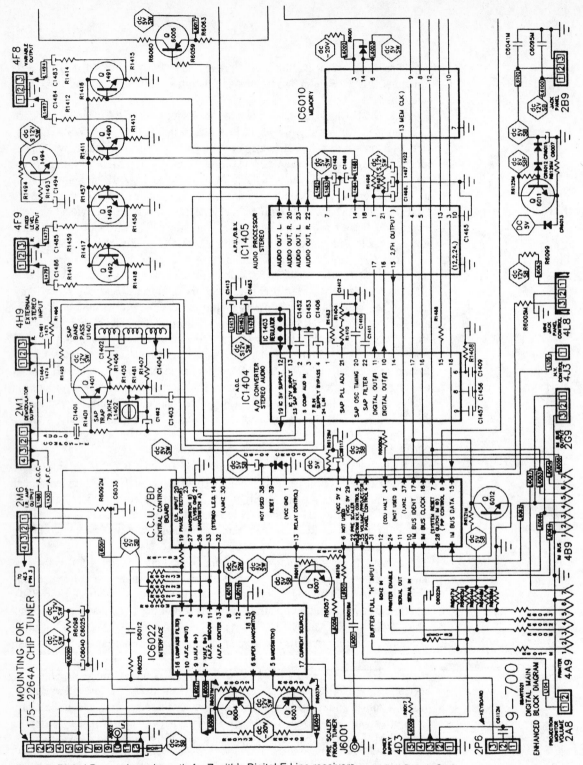

Fig. 5-7. Digital Processing schematic for Zenith's Digital E-Line receivers (courtesy Zenith Electronics Corp.).

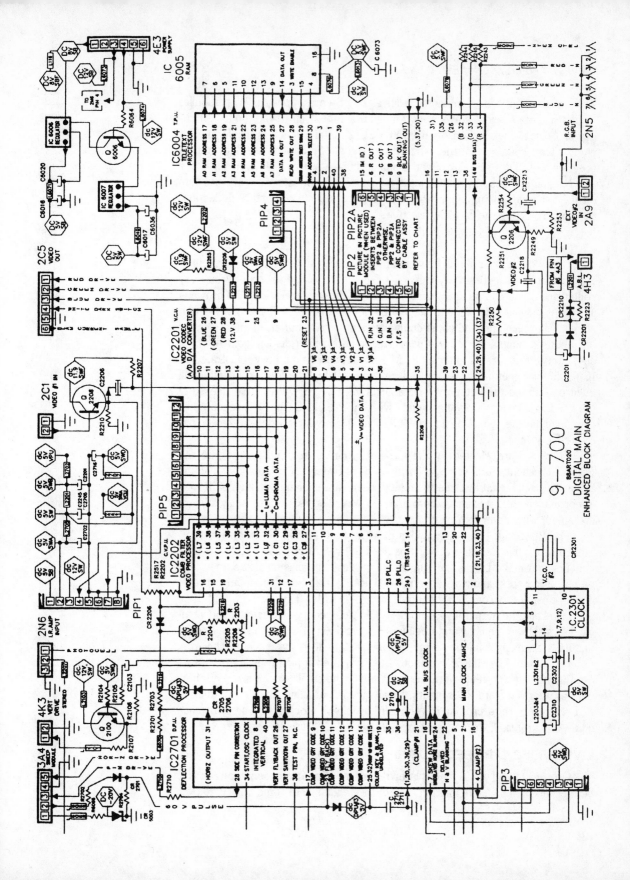

9-700
DIGITAL MAIN
ENHANCED BLOCK DIAGRAM

scan rates, filter time constants, vertical and S-corrections, E/W parabolic, widths and trapezoidal corrections for picture tubes, and sync-mode switch-overs. Analog video is first digitized in VCU 2133 and made available in 7-bit parallel Gray code so that the various sync and sweep options are delivered as programmed. Three-level key and blanking pulses are also derived for other allied digital processors. And there is a vertical flyback safety circuit that cuts off the vertical blanking pulse whenever problems occur in vertical deflection, thereby protecting the cathode ray tube.

TELETEXT

TPU 2732 is Zenith's IC 6004 and processes Level 1 Teletext information. Once again this is a 40-pin N-channel MOS circuit designed to accept 7-bit Gray code from VCU 2133 up to 8 stored pages, has ghost cancellation, and provides automatic language-dependent character selection. A block diagram of the unit shows data acquisition and ghost compensation, a RAM buffer, memory control unit, central timing, an IM bus interface, in addition to display control and a character generator. An external RAM (random access memory) illustrated as IC 6005, can be either a 64K × 1 bit or 16K × 1 bit, the former storing a full eight pages, and the latter two. Data acquisition begins with NTSC line 10 of the vertical blanking interval and ends with line 20, since line 21 is reserved as captioning for the deaf. Ghosts are filtered adequately if their delay times do not exceed 1 μsec.

Pages are preselected by the CCU microprocessor and loaded into the RAM. In readout, displays begin during vertical sweep at line 50 and end at line 242 so that text is not lost due to receiver overscan, each field being 262.5 lines in normal length. By loading appropriate RAM registers, it can display a list of stored contents upon command.

The PIP 2250, otherwise unidentified, is the ITT chip devoted to picture-in-picture for this Zenith receiver. Inserted between PIP2 and PIP2A, this IC contains four major blocks: horizontal and vertical anti-aliasing filters, input and output processors, and a dynamic RAM (DRAM) interface. The input converts digital luma and chroma information for the DRAMs. This data is read upon command and the output flows into the VCU video codec. Horizontal and verti-cal filters aid with peaking and frequency folding due to the 1:3 sampling rate for picture storage, in addition to interpolation filters to convert skew data from DPU 2543 into time shift information. But 80% of each picture is actually proc-essed, and only one-ninth of this is used due to the small picture size. PIP bor-ders are derived from RGB inputs. Small picture sampling times, according to ITT, are every third pixel and every third line, with storage in 5-bit luminance and 3-bit chroma. External memories of two DRAMS are 16 × 4K each.

S-VHS VIDEO TAPE RECORDERS

One might call these new player/recorders ''improved definition'' units, but because luminance resolution has been extended from some 3 MHz to over

5 MHz, the extended definition (EDTV) terminology is more appropriate.

This significant advance is credited primarily to engineers from the Victor Company of Japan (JVC) who, by moving the FM video (Fig. 5-8) format from 5.4 MHz to 7 MHz, increased carrier frequency to 2.6 MHz, and expanded frequency deviation from 1 to 1.6 MHz, have produced a separated luminance (Y) and color (C) picture that is easily today's basic forerunner of high definition television. Combined with almost dropout-proof new oxide tapes that can also be used for standard VHS recordings, and an increase of 2.6 MHz at white peak carrier, tape C/N responses are some 10 dB better than the originals as adjacent track crosstalk and picture stability respond comparably.

Recording tape speeds in SP and EP and 33.35 mm/sec and 11.12 mm/sec in ST-120 tape produce the usual 120 and 360 minutes, respectively, while chroma retains the original downconversion system of direct recording at a rate of 629 kHz. There are also broadband surface wave acoustical SAW filters installed in the tuners and a wideband Y/C separator that deserves an explanation.

This separator has a passband of 5 MHz, a 1H (horizontal line) CCD delay line and an adaptive filter to eliminate hanging dots and blurring of the vertically noncorrelated saturated chroma information. This latter is illustrated in Fig. 5-9 as a vertical transition detector consisting of two 3.58 MHz subcarrier bandpass filters connected to an AND circuit, whose outputs are routed through another bandpass filter and a chroma subcarrier trap. As illustrated, chroma is 1-line delayed and then continues on into the 3.58 MHz bandpass filter, while luminance has a charge-coupled device delay instead of the chroma glass delay line, is chroma filtered, and then passes into a 5 MHz low-pass filter for the indicated Y output.

For video displays that do not have Y/C inputs, Zenith furnishes a Y/C-to-RGB combiner that plugs into a printed circuit board on the rear of a TV jackpack (monitor). This 7- × 9-inch adapter connects to the receiver with fairly

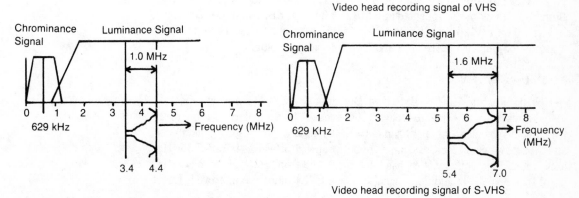

Fig. 5-8. How new VCRs increased their luminance bandpasses (courtesy IEEE Consumer Electronics and JVC).

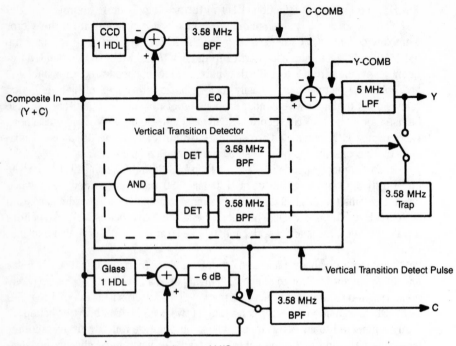

Fig. 5-9. The highly regarded comb filter and Y/C separator (courtesy IEEE Consumer Electronics and JVC).

long cables, and is said to deliver maximum S-VHS potential to the set. Later we suspect that the better receivers will all have Y/C inputs—Philips, for instance, already has this on some of their top models. This particular Y/C adapter features a 5-pin input, a video output, an off-on filter, in addition to color level and tint phase adjustment potentiometers. Its model number is S43A. Unfortunately we were not blessed with either theory of operation or a schematic, so the Y/C-to-RGB conversion process remains, for now, a small mystery. A pair of output cables and RGB sync connects to the TV receiver. Luminance response is specified at 5 MHz, +3 to −0 dB, and power consumption at 6.5 watts, 120 V ac. We also are told this converter will work for the Beta systems too—but it may require another input adapter plug.

ED-Beta by SONY

Next to actual HDTV, Extended Definition Beta (ED-Beta) at whatever viewing distance is going to deliver 500 lines of horizontal resolution (6.25 MHz) with very clean luma (Y) and chroma (C) separation. By using metal tape and moving luminance processing from between 4.4-5.6 MHz to 6.8-9.3 MHz, luminance frequency deviation has been increased from 1.2 MHz to 2.5 MHz, more than double. In addition, signal-to-noise (S/N) ratios are improved. Luma and chroma information are separately processed, resulting in lower crosscolor and dot crawl to the vanishing point, as well as excellent editing abilities with few picture effects.

Called high-band technology, a new, tilted Sendust Sputter head has also been developed to accommodate the metal tape's extended performance, which requires a much stronger magnetic field for recording, but exhibits considerable coercivity and offers very high density video recording. And because metal tape delivers playback with excellent S/N ratios, little noise reduction is needed, therefore, much better definition and resolution (fine detail) appears.

In basic color/luma separation, a conventional comb filter can often induce vertical and horizontal resolution losses, as well as diagonal resolution, dot crawl, and vertical color smear. To circumvent such problems, Sony has not only extended the luminance bandpass, but also introduced a new tape stabilizer for smoother playback and recording that relieves jitter and mottled colors, along with 20% smoother, ultra-treated surface tape. Consequently, medium and small amplitude signals now extend beyond 6 MHz rather than 3 MHz or less as before. This signal extension is considerably more linear (regular) due to a new linear and non-linear emphasis circuit that reduces amplitude fluctuations resulting from both tape and electronics.

A new comb filter had to be designed for further crosscolor and dot crawl removal, and this was done. Now, a new IC101 integrated charge coupled device (CCD) with 1H horizontal delay separates chroma about the 3.58 MHz center frequency, and the surrounding luminance spectrum components are switched at 10.7 MHz. In the form of a node-to-node bucket brigade, the faster-moving luminance is delayed a full line while chroma processing catches up. After these actions, a special correlation circuit detects and adjusts Y and C corrections in the signal, removing both cross color and dot interference. Following this, the Y comb filter reduces noise and adjusts for horizontal scan level differences. Y/C signals are now equalized, clamped, and continue on to preemphasis. This is nonlinear—as in conventional Beta—but reduced to ensure good video reproduction, FM modulation now takes place, the AFM 1.38 to 1.83 MHz frequencies are attenuated, Y/C is again mixed, AFM is also mixed, then amplified, and sent to the video head. At the same time, chroma is downconverted to 688 kHz.

For simple video or S-image outputs, video output from the correlation and dot interference section can be switched before the Y comb filter. Otherwise, S-image information bypasses the Y/C and correlation detector and dot removal circuits and either goes directly to an output or continues through the noise-reducing Y comb filter. Correlation, Sony explains, occurs when amplitude differences between delayed and non-delayed luma signals are 0, and Y/C are in identical phase. S-image inputs and outputs are pre-separated chroma and luminance information that allow maximum luma resolution and definition. When used with RGB monitors, a Y/C to RGB adapter is already available in Japan.

In playback, signals from the video head return through an AFM trap, are monitored by a dropout detector, limited, detected, low-pass filtered and de-emphasized before reaching the noise cancelling and dropout cancelling circuits. Following 6 dB amplification, luminance can proceed either to variable speed digital or directly to the Y and/or Y/C mixer for standard video output. Chroma initially travels much the same video head amplification and trap routes, but must have automatic color control as well as burst de-emphasis. The 688 kHz color downconverter information is upconverted to standard 3.579545 MHz in a re-converter II circuit, followed by the usual anti-harmonic bandpass filtering and the IC101 comb filter. Burst is further attenuated, and then returns to either the digital picture circuit or directly to either the S-image output or to the mixer for Y/C.

Tapes recorded on conventional Beta can be seen on ED-Beta with somewhat improved image quality due to Y/C. While ED-Beta metal tapes are *not usable* on standard Beta, they will not be automatically ejected. Standard cassettes, however, can record on ED-Beta but picture quality will suffer.

HI8 (8 mm)

While ED-Beta will be used more by professionals than the rank and file, Sony's brand new 8-millimeter equipment for cassettes, and especially camcorders, is a natural consumer item having almost as much video detail and outstanding digital sound.

Lightweight, (3.5 lbs.) with some 420,000 light sensing pixels and a precision $2/3$-inch charge-coupled device image pickup (instead of an ordinary glass receptor), it records up to two hours, operates down to 4 lux illumination, has an 8:1 variable speed power zoom lens, and variable shutter speeds from $1/60$ to $1/10,000$ second.

While ED-Beta boasts exceptional horizontal resolution of 500 lines, Hi8 produces 400 lines, which is some 70 lines more than perfect NTSC video reception. And while multichannel sound (MTS) is the best broadcasters can offer to date, Hi8's audio is totally digital for maximum bandpass and fidelity.

Improved horizontal resolution—usually 250 lines (about 3 MHz)—has been increased in the CCD-V99 to 400 lines (5 MHz) by the recently developed method of moving the luminance center frequency from 5 to 7 MHz, improving frequency deviation from 1.2 to 2 MHz and expanding signal-to-noise (S/N) ratios by both carrier frequency translation and the introduction of Hi8 metal E and metal P tapes. These are the metal evaporated and higher density magnetic material 8 mm tapes designed for low noise and excellent outputs versus the usual metal-particle tapes. Metal evaporated units are suitable for multiple recordings and editing, while P magnetic material is best for less expensive, but primarily original recordings.

Companion player-recorder to the CCD-V99 is the "jog and shuttle" remote-controlled EV-S900 with its double azimuth 4-head, narrow gap recording, sound track with less than 0.005% wow and flutter, synchro editing allowing variable high-speed forward and reverse search, slow motion, frame-by-frame advance, and clean freeze frame. Pulse-code modulation stereo digital sound can be recorded up to 24 hours in six 4-hour sound tracks having dual digital-to-analog (D/A) converters and correction circuits permitting a 90 dB dynamic range. On the back of this equipment you will find S-video (luminance Y and chroma C) inputs and outputs as well as dual sets of video and stereo jacks. The tuner receives 181 U/V/CATV channels, and has a programmable timer that accommodates 6 events in any 3-week period.

6
ATTC Setup and Testing

ALTHOUGH INITIAL TEST CONSIDERATIONS AND SOME REQUIREMENTS MAY change following various proponent system experiences, the basic procedures have been agreed upon and test equipment is being or already has been set up for a good many final examinations. The Advanced Television Test Center (ATTC), under the executive direction of Peter Fannon, with expert support from Chief Scientist Charles W. Rhodes, formerly of Tektronix and Philips Laboratories in New York, and National Association of Broadcasters communications specialist E. Benjamin Crutchfield as Program Officer, and Edmund A. Williams as Manager, Transmission and Propagation Engineering, began operations in 1989 and expect to complete most of the work within two years.

ATTC was formed to test and evaluate advanced television technologies in cooperation with the Advanced Television Systems Committee and the Federal Communications Commission's Special Advisory Panel on advanced television. Sponsors, who have contributed $3.5 million include NAB, CBS, Inc., the Association of Independent Television Stations, Association of Maximum Service Telecasters, Capital Cities/ABC, Inc., the National Broadcasting Co., Inc., and the Public Broadcasting Service. Systems submitted will be evaluated by this group and the results reported to the Advanced Television Systems Committee that will make its considered recommendations to the Federal Communications Commission, and probably the U.S. Congress. Such could occur late 1991, depending on unanticipated test problems, late system deliveries, and/or the finalizing of reports. The FCC should return some sort of decision by late 1991 or early 1992.

Meanwhile, cable television (CATV) and satellite transmitters will probably be experimenting on their own. Channel bandwidths in these communications media are not nearly as restrictive as in 6 MHz terrestrial broadcasting. However, because consumer-developed receivers will not be readily available to the public, another several years will have to pass before HDTV can become commercially practical—and then prices on the first receivers could easily surpass $2,000 or more, especially those with large, widescreen 16:9 aspect ratio picture tubes. Mass production can reduce these prices sharply, but HDTV will certainly not be cheap now or in the foreseeable future.

TEST PLANS

ATTC testing takes place at radio frequencies (RF) rather than audio/video baseband, and proponents have been required to furnish their own modulators and demodulators. System encoder inputs were specified as red, blue, and green (RGB) and audio, with modulator outputs at an i-f of approximately 45.75 MHz, which is the conventional video carrier frequency nominally seen at 50% on receiver swept response curves. And in the event you wish a reminder of i-f and trap frequencies delivered (passed) by NTSC tuners here they are, all in megahertz:

I-Fs	TRAPS
41.25 sound carrier	35.25 upper adjacent sound trap
41.67 lower chroma sideband	39.75 upper adjacent picture carrier
42.17 chroma subcarrier	41.25 sound carrier (partial)
42.67 upper chroma sideband	
45.75 video carrier	47.25 lower adjacent sound carrier
Note: better U.S. receivers contain separate SAWs for video and audio.	Note: most trapping now executed by surface wave (SAW) acoustical filters between tuners and i-fs.

A typical receiver i-f response curve with carrier and trap frequencies is illustrated in Fig. 6-1. SAW filters remove undesirable (trap) frequencies with precision passbands and sharp skirts. SAW losses are usually compensated by single-stage preamplifiers, a practice we would expect to continue in HDTV as 6 MHz channel bandpasses remain the same whether single or augmented channels in terrestrial broadcasting, and probably the same for CATV. Satellite transponder bandwidths between 36 and 72 MHz could go "brute force" without signal separation since their transmissions are frequency modulated (FM) and satellite receivers might easily be modified to provide the rest, including Y-C and/or RGB outputs. They already have broadband composite and video/audio ports available for VCRs, monitors, and descramblers. Only a few extra circuits would be needed for the remainder. Ordinary TV receivers, however, would only produce standard 4:3 aspect and 3-4 MHz resolution pictures, even with good comb filters.

Modulation is/was permitted either at VHF or UHF, with proponents supplying all equipment. Two specific signals at RF are generated, one at NTSC

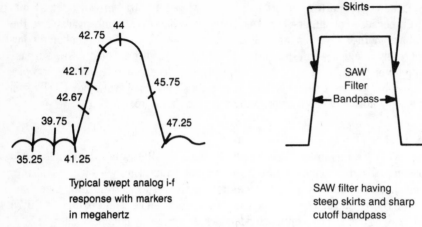

Typical swept analog i-f
response with markers
in megahertz

SAW filter having
steep skirts and sharp
cutoff bandpass

Fig. 6-1. Typical swept i-f response curve and SAW filter that eliminates tuned circuits between i-fs and tuners.

and the second at ATS. The first calibrates the ATTC "test bed"—an example is illustrated in Fig. 6-2—and then operates as a prime signal for interference testing. Four NTSC modulators serve as interference simulators for co-channel, adjacent channel, and the UHF taboos, all of which belong to the Test Center. And to maintain minimum interference and noise, the test bed operates at a high signal level.

As for signal processing, an NTSC reference reacts to measured noise, interference and multipath signals from each proponent system. A multipath generator delivers leading or lagging secondary images as well as three ghosts at selected levels, plus delays. Configurations include bypass, multipath or termination modes.

The third portion of the test facility supplies further amplification, a group of Electronic Industries Association (EIA) receivers, a spectrum analyzer, two sets of NTSC and ATS demodulators, and a pair of NTSC and ATS decoders delivering final RGB and audio outputs.

Receivers will test co-channel, adjacent, and taboo channel rejection from ATS in "just perceptible" tests. NTSC demodulators supplied by the testing facility are precision units even though each system offeror will be required to furnish his own as well as a decoder, especially if the output is a compatible NTSC signal.

In general, that's the ATTC setup to be followed throughout the various system examinations. As testing continues, some details could be altered, but the main principles remain in effect throughout. Naturally, as certain susceptibilities become apparent, more attention is paid to them, with more or less emphasis on obvious or special problems. As in all competition, some pass and others fail, much depending on good engineering that meets FCC guidelines

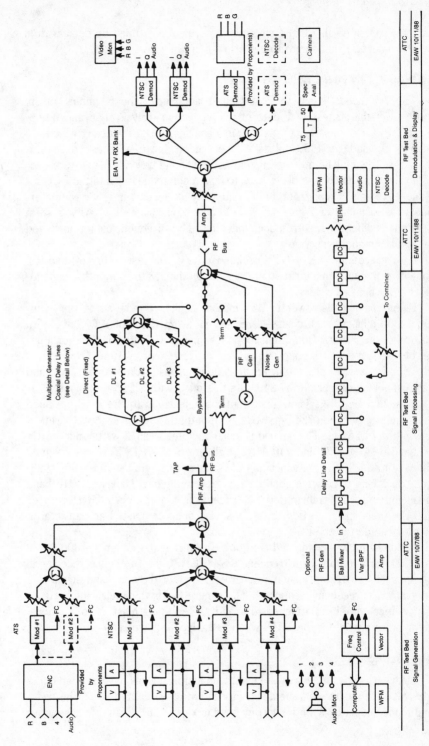

Fig. 6-2. Preliminary ATTC test bed with modulators, delay lines, and detectors designed for HDTV system testing (courtesy ATTC).

and industry approval. Unique circumstances, naturally, must be dealt with in their own context, but all are to perform within the rules.

PROPAGATION ANALYSIS

The Advanced Television Test Center announced the beginning of six months over-the-air testing to determine propagation analysis of advanced television (ATV) transmission possibilities in Washington, D.C.—testing which will probably conclude in 1990 unless it is extended due to unforeseen circumstances. Following laboratory testing in mid-1990 and beyond, definitive broadcast evaluation should take place prior to final recommendations to the FCC for all likely systems. Propagation tests, underway in December 1988, will determine the feasibility of 2-channel HDTV operations as well as ATV system potentials in different spectrum locations, including "frequencies not now used for regular broadcast service."

As of December 7th, HDTV preliminary open air testing began, and will probably continue in one form or another for another 18 to 24 months, depending on proponent survivors.

Using channel 9's tower in Washington, D.C., over-the-air testing commenced on experimental channels 58 and 59 at an EIRP of 1700-2500 with approval of the Federal Communications Commission. Tests continue for a period of one year from the operational date of December 4, 1988. Transmitter output power is 700 watts, sufficient to conduct propagation tests on single and dual-channel operations during 24 hours operation with NTSC emissions.

The Advanced Television Test Center's application to the FCC specified one frequency between 734-746 MHz at an output power of 1778 watts, and a bandwidth of 12 MHz. The second channel operates at 2512 watts on a 6 MHz passband between 734 and 740 MHz. Call letters WWHD-TV were assigned originally to the National Association of Broadcasters on December 30, 1986. Here, dual band tests were conducted to determine how UHF and VHF band signals differed when transmitted under identical circumstances. By December 1989 the preliminary portion of this test had been completed and an interim report forwarded to the FCC.

Field testing was conducted at both UHF (ultra high frequency—300-3,000 MHz) and SHF (super high frequency, 3-30 GHz), with a mobile receiver in operation to "analyze propagation characteristics of the wideband signal at a wide variety of receiving locations." The vehicle had both circular and horizontally polarized antennas attached to a 30-ft. telescoping mast. Two additional antennas were used for the SHF 2.5 and 12 GHz frequencies.

UHF Testing

Dual horn-type antennas were used with identical gains but polarized circularly and horizontally with effective radiated powers of 32.5 dBW in the wideband mode and 34 dBW in the narrow band mode, or 2512 watts, at respective antenna input powers of 25.5 dBW and 27 dBW. Both antennas were mounted adjacent to one another at the 280-ft level, aimed at 235° in a southeasterly direction, and covered the area between due south and due west. During the tests, channel 58 was identified for a 30-second period with call letters, location, and telephone number via an NTSC character generator each half hour. In dual band operation, the carrier is positioned at 740 MHz between the two and emits double sideband.

Program material for these tests included video test signals as well as programming from channel 9 (WUSA-TV). Differences between Ch. 9 and Ch. 58 emissions were recorded separately to detect differences, including normally programmed audio.

SHF Testing

Both 2.5 and 12 GHz SHF frequencies were specifically selected, the lower frequency is now occupied by the Instructional Television Fixed Service (ITFS), while the higher spectrum may "support" a broadcast possibility for HDTV. Especially the ATTC wants to check signal strengths, multipath, and phase shifts of *both* AM and FM modulation formats. A 200 W amplifier was installed adjacent to the wide-beam antenna on Ch. 9 so line losses could be "eliminated" and considerably more power radiated. A center frequency of 12.450 GHz was used. This system operates at the same time as UHF radiations and field measurements were taken.

Projected 2.5 GHz transmissions at wideband operate at an effective radiated power of 2,512 W, and the 12 GHz frequency delivers an EIRP of 2,951 W. All antennas for both UHF and SHF were mounted approximately 280 ft above ground, with no buildings or other structures within 250 ft of the main antenna beams.

Test results will probably not be made available until well after this book is published. We have no idea whether the FCC would actually allow commercial terrestrial GHz transmissions at these frequencies, especially since K-band satellite downlinks are already authorized at 11.7-12.2 and 12.2-12.7 GHz. Perhaps such tests will determine the possibilities of any direct interference. More Tx/Rx problems we do not need, especially with the advent of a brand new service.

SYSTEM CONSIDERATIONS

With specific testing already scheduled, you might also like to know some of the Committee and ATTC concerns prior to hardware receipt in order to evaluate future commercial-consumer electronics sure to appear in the next several years.

Physical aspects of similar equipments will become smaller due to large-scale integration in solid state ICs, but programming into microcomputers, microprocessors, EEPROMs, RAMs, ROMs, and the like will more than quadruple as silicon real estate allows vastly more complexity on monolithic chips.

Although first-rate engineering is expensive, silicon/sand is the second most common element on earth, and IC prices always decline substantially after primary market introduction. Consequently, initial products will usually appear in high dollars, but decline rapidly in price as competition and production force them down. We are therefore deliberately adding this section so that readers can absorb searching engineering questions on just what constitutes any of these proposed HDTV or EDTV systems and how they operate. Later, you may want to ask a few when acquiring a system or just a receiver. Many points continue to remain pertinent as the remarkable world of video in business and entertainment grows.

Descriptive Questions

What are system basics and transmission format? Is it NTSC compatible, 525-line sequential, composite or time-multiplexed components, the number of separate or adjacent channels? What are the composite baseband signals at decoder output(s), and is/are there one or two such as in quadrature modulation? Other encoders may supply no clearly separate baseband but have multiple signal components that directly modulate the carrier before transmission.

Indicate bandwidths and amplitudes of all video signals, and how are they processed—whether stretched, compressed, or folded before being encoded—as well as baseband video structure, identifying each in the main channel, and any in the augmentation channel? Where frequency multiplication occurs, name the subcarrier frequencies, their relationship to other signal characteristics, and bandwidth/modulation for each.

Where there are time-multiplexed components, ATTC and ATS wanted to know if they were line-related or multiplexed into horizontal blanking intervals and if related or unrelated to standard TV formats? Any and all video compression was also of interest, signal bandwidths before and after compression, and the times and locations.

Certain subsampling and filtering techniques often reduce signal bandwidths, and their operations are important. Especially the types, bandwidths and rolloffs of pre- or post-filtering and the field or line sampling accompanying them. And what about digital processing in the encoder/decoder stages?

There is also need to know sampling frequencies, the number of samples per line for compression or multiplexing of both luminance and chroma, or any other equivalent signals.

The composition of video information relative to luma and chroma components undergoing compression, scaling, bandwidths, and transmission axes at strategic points throughout the system are required. Then, too, color primaries and reference white specifications at the camera may also be needed to allow for future upgrading or improvements.

Sync

The horizontal line scan time in any NTSC signal is 63.5 μsec, of which 11.1 μsec constitutes the blanking interval and the 4.76 μsec horizontal sync pulse. In each vertical field, there are 21 lines appearing in 1.33 msec, accommodating six vertical sync pulses preceded and followed by equalizing pulses, all of which are keyed to the horizontal line period (H). The video portion of any H line is, therefore, 52.4 μsec. The first 9 lines of the vertical blanking interval carry vertical sync, while lines 10 through 20 are authorized for Teletext, line 19 for the VIRS color signal, and line 21 is set aside for hard-of-hearing captioning.

This shows that 17.48% of horizontal scan and 7.98% of each vertical field (two fields make a frame) are reserved for horizontal and vertical blanking. Under FCC regulations, these will all continue so that ordinary TV receivers can operate at NTSC without EDTV or HDTV interference. But there may be auxiliary signals also delivered among the NTSC periods and special transmissions, including sync, in any augmented channels. So this information is most certainly required. Also, what correlation and amplitudes connect secondary and primary sync transmissions, especially in noise and secondary image conditions. And what about undesirable artifacts resulting from sync crosstalk in the picture?

Resolution

In the end, what's of prime interest is a wideband, broad aspect picture with virtually no glitches or intrusions called artifacts. But all lines, pixels, images, still scenes, and movements should also be clean and received in much the same mode, color, and luminance (brightness and detail) as transmitted.

Resolution is measured in terms of *spacial* and *temporal*; *spacial* being defined in numbers of lines per picture height, and *temporal* as the number of fields necessary for maximum receiver horizontal and vertical resolution. While the latter is primarily subjective, dealing with motion and time, spacial can be directly calculated by multiplying the active line time by the compressed video bandwidth times the inverse of the screen's aspect ratio. They are easily remembered as space for spatial, and time for temporal in plain English.

Otherwise, luminance is normally judged by the number of lines or megahertz (MHz) it can resolve from a calibrated test pattern or multiburst generator, loosely characterizing the relationship as 80 lines/MHz. But since most television receivers are not tied to high definition cameras, we prefer to work with megahertz rather than lines wherever possible, especially when these equipments have monitor "jackpacks" for audio-video inputs and bandpasses that usually exceed 6-7 MHz. Composite video/audio baseband outputs from satellite receivers are often 8 MHz.

Colorimity specifications and transfer characteristics are also of considerable importance throughout any television system, especially any compression, scaling, passbands, and transmission axes.

Test Schedules Have Been Set

While there may be no such thing as "absolute," the FCC Advisory Committee on Advanced Television Service has now set 1990/1991 dates for proponent system testing, beginning on Friday, May 25, 1990, and ending on Tuesday, September 3, 1991—a period of some 16 months from start to finish. Ten days will be allowed for "move-in" and 5 days for "move-out" during the beginning and ending of each test period.

A reservation fee of $25,000 for ATV System Access Period is required which will be credited towards assessed test fees which can range from $175,000 for first system testing to $300,000 for a second, subsequent or synthesized system, and/or retesting, in addition to a penalty of $12,000/day for testing needed beyond the assigned test period that may not exceed a total of 30 days, regardless.

In addition to these charges, there can also be audio and field testing fees assessed, neither of which was tabulated by December 1989.

Proponents scheduled for system tests in 1990 are:

Faroudja	May 25 to July 16
PSI	July 17 to Sept. 4
Sarnoff (ACTV-1)	Sept. 5 to Oct. 23
NHK (Narrow MUSE)	Oct. 24 to Dec. 12
NHK (MUSE 6)	Dec. 13 to Feb. 8, 1991

Proponents scheduled for system tests in 1991 are:

Zenith	Feb. 11 to Apr. 1
Sarnoff (ACTV-II)	Apr. 2 to May 20
Philips	May 21 to July 11
MIT	July 12 to Sept. 3

As you can readily see, there are only 9 separate tests scheduled (although Japan Broadcasting may reconsider) out of a total of 17 who expressed initial interest. And of these nine, NHK, and Sarnoff Princeton Laboratories have two slots each, so there are now only 7 proponents actually remaining, although no audio has yet been scheduled.

Further, should Thomson and Philips join forces as rumored, this combination could share Philips' reserved slot in the May-July 1991 period, or request another time for testing, if available. This new system, it is said, would be much like Zenith's dual channel NTSC and companded simulcast arrangement, with HDTV compressed from 30 MHz to 6 MHz, then expanded again to 30 MHz at the receiver. Could Zenith become the third partner?

Many Problems

The biggest problem of all involves the readiness of various proponents having equipment assembled and operating as each "due date" arrives.

With this in mind, Systems Subcommittee Working Party 1 intends to make on-site preliminary inspections of each system well in advance of actual testing to be sure all is ready, willing, and able. WP1 also has required a T-90 (days) system update, including a complete technical description of the several systems by December 31, 1989.

This announcement did not elicit a ringing endorsement from some system proponents who would rather work solely with the testing group which still hasn't full system parameters or RF information, both of which are totally necessary to properly configure the local ATTC test bed. Nor is any specific audio or modulation system yet called out in any system.

WP1 is the certifying working party that notifies the Advanced Television Test Center through Systems Evaluation and Testing (WP2) if and when a proponent system is ready for full examination. Those failing either technical qualifications or test slot readiness could either trade with another proponent for a later test time or actually be dropped from the program. Contractual time for everyone is now becoming crucial and testing must be complete before the ATSC authorization expires in November 1991.

Once the Federal Communications Commission finally selects one or more high definition systems for terrestrial service and receiving equipment becomes available, broadcast stations must modify or add to their own towers and exciters to accommodate those extra channels which seem certain to be required. An on-going survey suggests that many TV antennas are 25 to 30 years old and won't support additional weight and wind resistance, others are already erected in cramped quarters, and independent stations probably won't have the projected several million dollars needed for HDTV changeover.

Apparently, broadcast network TV with captive stations may well win the day, along with CATV and, possibly, Ma Bell as their fiberoptics home TV phone experiments continue in Florida and possibly elsewhere.

ATS SURVEY QUESTIONNAIRE

For further specific inquiry, the following questionnaire was sent to all proponents to determine specifications, capabilities, limitations, and extent of each particular EDTV or HDTV equipment. Originator, the Implementation Subcommittee of the Advisory Committee on Advanced Television Service (ATS), wanted "to examine transition scenarios for proposed systems." Translated, this means they wanted to have a look-see into each proposal.

The ATS Survey Questionnaire

1. Which category correctly describes your proposed system?

    ____NTSC with added processing and no added information
    ____NTSC with added processing and information, bandwidth unchanged
    ____NTSC with added processing and information, bandwidth increased
    ____NTSC with added processing and information, bandwidth initially unchanged, but extendable
    ____Non-NTSC, bandwidth unchanged
    ____Non-NTSC, bandwidth increased
    ____Non-NTSC, bandwidth initially unchanged, but extendable

2. For which form(s) of distribution is your system intended?

    ____Terrestrial Broadcast
    ____Multichannel, Multipoint Distribution Service
    ____Cable Satellite Distribution to Broadcast Stations, and Cable Head Ends
    ____Satellite Broadcast Direct-to-Home
    ____VCR/Disc

3a. With which production signal format(s) is your system designed to work?

    ____NTSC
    ____Component Video (specify) _____
    ____Other (specify) _____

3b. Does your system require digital as opposed to analog implementation in the production environment?

    Yes _____ No _____

3c. With which production scan rate(s) is your system designed to work?

    ____525/59.94/Interlace
    ____525/59.94/Progressive
    ____1050/59.94/Interlace
    ____1050/59.94/Progressive
    ____1125/60.00/Interlace
    ____Other (Specify) _____

4. What is required to interface your system to each of the production signal formats and scan rates you checked in questions 3a., 3b., and 3c. above? Attach a separate sheet.

5a. Can your system be viewed on existing TV sets without the addition of any new equipment in the home?

Yes _____ No _____

5b. Will any of the picture parameters be different from current NTSC?

Yes _____ No _____

If yes, describe on a separate sheet which ones and how they will be different.

5c. Rate the quality of reception on existing TV sets versus NTSC signals using the 7-point comparison scale from CCIR Recommendation 500-3, Table II, as follows:

____ +3 Much better
____ +2 Better
____ +1 Slightly better
____ 0 The same
____ −1 Slightly worse
____ −2 Worse
____ −3 Much worse

6. What is required to interface your proposed signal format to existing TV sets?

____ Nothing
____ Set Top Converter. If yes, describe on a separate sheet.
____ Special Connector. If yes, define on a separate sheet.
____ Other (specify on a separate sheet)

7. What is your recommendation to receiver/monitor suppliers with regard to an interface that would work with your system, conventional NTSC, and possibly other systems?

____ E.I.A. Multiport (Interim Standard is 15 as revised to include Y/C and pay-per-view)
____ Wideband RGB
____ Wideband Y/C
____ Other (specify on a separate sheet)

8a. At which stage of development is your system at present?

_____Paper Design
_____Computer Simulation
_____Laboratory Breadboard
_____Engineering Prototype
_____Production Prototype

8b. When do you estimate each of the remaining stages listed in 8a. will be reached? Give the quarter and year.

Paper Design _____

Computer Simulation _____

Laboratory Breadboard _____

Engineering Prototype _____

Production Prototype _____

9a. Is LSI development necessary for production of your system?

Yes _____ No _____

9b. If yes to 9a., what is the status of this development? Answer on a separate sheet. Include the approximate gate count of any unique chips required?

10. Are there other critical devices necessary for your system which require development?

Yes _____ No _____

If yes, what are they and what is their status? Answer on a separate sheet.

11a. Does your system contemplate more than one phase of improvements?

Yes _____ No _____

11b. If yes, over what time frame will the additional improvements be available?

_____Immediately
_____After FCC regulatory and allocation action
_____After _____ years from system _____ implementation or _____ adoption
_____Other (Specify on a separate sheet.)

11c. Do the subsequent improvements require additional spectrum space?

Yes _____ No _____

If yes, indicate on a separate sheet the amount of spectrum, when it will be required relative to the stage(s) of improvement(s), and its location in the spectrum.

11d. What are your recommendations for receiver/monitor interfaces for any later phases?

11e. What changes will be required in the production system or the interface(s) to it?

12a. Is conditional access a feature of your system?

Yes _____ No _____

12b. If yes on 12a., is it compatible with conventional or existing conditional access systems.

Yes _____ No _____

13. Does your system provide for the continuation of ancillary services that are presently available? Check all that are retained.

____Teletext or other data services
____Vertical interval test signals
____Captioning for the hearing impaired
____Conditional access control signals
____BTSC stereo sound channels including the Separate Audio Program (SAP) channel and the Professional channel

14. Does your system provide additional ancillary services that should be considered in planning transition scenarios?

Yes _____ No _____

If yes, describe on a separate sheet.

15. What new or modified test equipment, if any, will be required to implement your proposed system (demods, waveform monitors, spectrum analyzer, tunable filters, etc.)? Describe on a separate sheet.

Please answer the following questions if your system is intended for terrestrial broadcast:

16a. Does your system require regulatory approval or changes for its initial implementation?

Yes _____ No _____

If yes, describe on a separate sheet what aspects of your system require such approval.

16b. Is further regulatory approval or regulatory change required for implementation of phased improvements?

Yes _____ No _____

If yes, describe on a separate sheet what aspects of your system require such approval.

17. Is more than 6 MHz required?

(a) Initially?

Yes _____ No _____ If yes, how much? _____MHz.

(b) With phased-in improvements?

Yes _____ No _____ If yes, how much? _____MHz.

18. If additional frequency spectrum is needed, will it be used to enhance the existing service, or will it be used to provide a new service (i.e. for simulcasting)?

Enhancement _____ Simulcasting _____

19. If your system uses an augmentation channel:

(a) Does the augmentation channel need to be contiguous?

Yes _____ No _____

(b) If the augmentation does not need to be contiguous, is there any limitation on separation between channels?

Yes _____ No _____ If yes, explain on a separate sheet.

(c) Are there special performance requirements on any parameters, e.g. group delay, which must be maintained between the two channels?

Yes _____ No _____ If yes, explain on a separate sheet.

20. Describe, on a separate sheet, the robustness of your system with regard to:

   Sensitivity to Multipath
   Protection Ratios Required (co-channel)
   UHF Taboos

21. Can existing point-to-point microwave transmission systems (e.g. Studio-to-Transmitter Links) accommodate your system?

   Yes _____ No _____

   If yes, with _____ or without _____ modifications?
   If no, explain why, or if yes with modifications, describe on a separate sheet.

22. On a separate sheet, describe the impact of your system on service provided by present broadcast translators. Include considerations such as the expected levels of performance relative to NTSC performance, the use of remodulating translators, the use of heterodyne translators, and the like.

23. Is there anything else about your proposed system which might be uniquely affected by broadcast transmitter and antenna systems?

   Yes _____ No _____ If yes, describe on a separate sheet.

Please answer the following questions if your system is intended for cable:

24. Is more than 6 MHz required?

   (a) Initially?

   Yes _____ No _____ If yes, how much? _____MHz.

(b) With phased-in improvements?

Yes _____ No _____ If yes, how much? _____MHz.

25. If additional frequency spectrum is needed, will it be used to enhance the existing service, or will it be used to provide a new service (i.e. for simulcasting)?

    Enhancement _____ Simulcasting _____

26. If your system uses an augmentation channel:

    (a) Does the augmentation channel need to be contiguous?

        Yes _____ No _____

    (b) If the augmentation does not need to be contiguous, is there any limitation on separation between channels?

        Yes _____ No _____ If yes, explain on a separate sheet.

    (c) What is the maximum difference in signal level between the main channel and the augmentation channel, measured at the subscriber location, that can be tolerated by your system?

        _____ dB

    (d) Are there special performance requirements on any parameters, e.g. group delay, which must be maintained between the two channels?

        Yes _____ No _____ If yes, explain on a separate sheet.

27. Can existing point-to-point microwave transmission systems (e.g. Community Antenna Relay Service) accommodate your system?

    Yes _____ No _____

    If yes, with _____ or without _____ modifications?
    If no, explain why, or if yes with modifications, describe on a separate sheet.

28. Does your system require a level of performance at cable headends and cable distribution systems that is currently in place:

____On virtually all existing cable systems?

____On newer well-maintained cable systems only?

____On hardly any existing cable systems without extensive modifications? Give details of such modifications on a separate sheet.

29. What are the effects of passing your signal through existing cable set top converters, both RF and baseband? Is the resultant quality (check one in each column):

RF	*Baseband*	
____	____	The same as present NTSC?
____	____	Degraded? How much? _____
____	____	Improved? How much? _____
____	____	Requires replacement converter?

30. How would your system be affected by transmission through broadband, as opposed to channelized, CATV microwave systems?

____No effect

____Some degradation. Explain on a separate sheet.

____Would not operate on broadband microwave without modification. Explain on a separate sheet.

31. What effect will reflections in a CATV system have on your proposed system? Give details on a separate sheet.

32. What is the minimum carrier-to-noise ratio your system requires on CATV systems to meet your proposed system performance requirements?

33a. Can FM modulation techniques be utilized with your proposed system for signal transportation purposes (e.g. fiberoptic transmission, microwave transmission, satellite transmission)?

Yes _____ No _____

33b. If yes, what FM bandwidth is needed? _____MHz.

34. Will there be any new requirements on radiated signal ingress for CATV systems that carry your proposed system signals?

Yes _____ No _____ If yes, describe on a separate sheet.

35. What new requirements will be placed on equipment such as video switchers, RF combiners, and broadband amplifiers (e.g. Composite Triple Beat, Cross Modulation, Carrier-to-Noise Ratio, Group Delay)? Describe on a separate sheet.

36. On a separate sheet, describe how your proposed system would be affected by:

    (a) RF scrambling techniques (e.g. sync suppression)
    (b) Baseband scrambling techniques (e.g. random video inversion and sync suppression in combination)

37. Is there anything else about your proposed system that might be uniquely affected by transmission through broadband CATV/MATV systems?

    Yes _____ No _____ If yes, describe on a separate sheet.

Please answer the following questions if your system is intended for satellite transmission:

38. For the type of service contemplated, what are the requirements for:

Transponder Bandwidth	___ MHz
Transponder Power	___ W
Related earth station performance (G/T)	___ dB/K

39. What carrier-to-noise ratio is required for your system to meet its stated objectives?

    _____ dB

40. Do you foresee using one or two satellite transponders per advanced television channel:

	1 transponder	2 transponders
(a) Initially	_____	_____
(b) With phased-in improvements?	_____	_____

7
HDTV-CATV Considerations

THE CABLE TELEVISION INDUSTRY (CATV), ALTHOUGH SOMEWHAT AMBIVALENT AS to standards and specific procedures early in 1989, cautions that there are three distinct components in the HDTV approach:

- Production
- Transmission
- Display

Members cite the "already . . . very high level of acceptance in this country and in other parts of the world" of Japan's 1125/60 format as a production standard. But they said, the ATS committee "recognized that additional formats also might be adopted." And they added that one or more transmission formats becomes much more complicated, especially in serving the nation's 90 million some TV households during transition times from NTSC to HDTV. The National Cable Television Association claims that 53% of U.S. homes receive broadcast television programming via cable. The Association wants an advanced TV system to offer both high quality broadcasts as well as cable re-transmission of such high quality programming. Cable spokesmen declare cable and other media have "certain characteristics and capabilities" that could offer "unique" HDTV services. And they recall the FCC announcement not to slow down introductions of advanced TV systems by the non-broadcasters. They also urge Congress "not to impede cable's ability to provide the best it can deliver." And while they admit that standardization may be required for HDTV broadcast transmissions, "alternative video distribution media should be left free to maximize their particular technical capabilities in the marketplace."

In an ideal world, CATV reasons, a single, uniform transmission standard for all media such as broadcast, cable, DBS (direct broadcast satellites) and video cassette recorders would be appropriate. But peculiar characteristics of each medium could necessitate multiple transmission standards—a supposition now being studied to "interconnect different transmission formats to a single receiver display mode" to avoid consumer confrontation with a number of incompatible standards.

There is a consensus among industry on testing and "inter-operability," says CATV, but what's needed later is an analysis of the various issues as HDTV systems begin to operate, as well as a successful transition from NTSC to HDTV for all video delivery media.

Overall economic and policy questions must be founded on technical issue resolutions and "comparative system analyses." Therefore, performance claims, transmission capabilities and spectrum demands of the various systems have to be substantiated by real hardware testing prior to any standards resolutions. And testing now is only limited by hardware availability. Cable Laboratories, Inc., is now actively working with the Advanced Technology Test Center (ATTC) in preparation for testing alternative ATV systems for cable "carriage." (Cable Labs) was created in May 1988 by cable operators for research and development to work with ATTC in developing new technologies suitable to CATV.

CATV interests do not want the FCC "to narrow the scope of its ATV inquiry" by setting early spectrum parameters, believing that spectrum issues and related technical proceedings "are inextricably bound up in the selection of standards . . . and there is no need to decide these issues on the basis of thin information."

The CATV group could also see a definitive need for subjective testing by viewers of added video resolution and other enhancements. To that end, the ATTC and Cable Labs are now working on a psychophysical test plan. Already a number of recommendations have been made as to the number and length of test segments, screening/ranging, viewing room conditions, procedures, display type and size, and a psychoacoustic test plan by SS WP-2. Subjective tests, it's said, could be conducted with laboratory electronic tests, or test signals might be recorded and run at a different time and location. ATTC is investigating the feasibility of video tape recording each ATV system so that subjective tests can be undertaken later at their own and other facilities—they say it could be cost-effective and offer significant advantages. Subjective testing is expected to involve public citizens with normal visual and auditory acuity over a period of several hours, with results tabulated.

The final test phase after the lab and subjective testing involves over-the-air broadcasting on surviving systems to verify predicted performance. This is to be done with the aid of cooperating broadcast facilities. Meanwhile, preliminary propagation testing continued throughout the summer of 1989 to foresee any significant problems, especially with those systems that require more than the standard 6 MHz system bandwidth. NBC and CBS are both loaning equipment, including a field measurement van from CBS.

Over-the-air testing has been underway from WUSA-TV, channel 9 in Washington, D.C. Group W. Broadcasting's WJZ-TV Baltimore, channel 13, and the Wisconsin Educational Broadcast Board have offered their TV and microwave facilities. Other offers are expected, and the ATTC will welcome such cooperation, especially for certain test equipment, loan of technical personnel, and other project needs as testing moves into evaluation stages. At the same time, Canada is helping with equipment and the Canadian Communications Research Center is working on software for the project.

While all the above doesn't necessarily involve cable, it does offer a general view of testing now underway and planned. And to that end, the Test Center has listed the names, status, hardware, category, and public demonstrations planned in 1989, most of which occurred at the National Association of Broadcaster's convention in Las Vegas that spring. Laboratory testing, as previously stated, began in the Washington area and possibly other selected sites in fall 1989. A preliminary draft of signals and tests for terrestrial communications is illustrated in Fig. 7-1.

FORMATS AND PRODUCT

It is considered that CATV system bandwidth will be available as required and will need no distribution system upgrades—this is predicated on the assumption that any systems now below standard will either upgrade or eliminate "under-utilized" channel programming by the time HDTV is ready. If and when upgrading is needed, a CATV operator would probably increase bandwidth from 300 to 450 MHz to accommodate high definition channels between 9 and 12 MHz wide. If signals are received by satellite, local NTSC conversion will be offered for those with standard receivers. As to local terrestrial service, cable will provide the same format as transmitted initially, if such a format is suitable for CATV re-transmission without signal impairment. The initial marketing model indicates a 3% HDTV penetration fairly quickly.

Primary equipment costs in CATV systems are for distribution trunk amplifiers, cable and converter/decoder subscriber equipment, but installation and maintenance are also significant factors. Consequently, total rebuilding of a cable TV system would represent a considerable financial burden to its owner/ operator. So HDTV systems that will not need heavy additional investments for cable owners are obviously preferred.

While the initial economic impact of HDTV on CATV is expected to be relatively small because of the "slow penetration of HDTV receivers," cable TV will be ultimately "impacted," especially with respect to so-called premium TV services. And the installation of additional expensive decoders is not relished. CATV operations must now look carefully at any and all extra technical parameters as well as those delivering an adequate NTSC signal. Equipment cost considerations are uppermost in anticipated HDTV delivery.

The National Cable Television Association anticipates the possible adoption of one or more studio production standards and the selection of media-specific

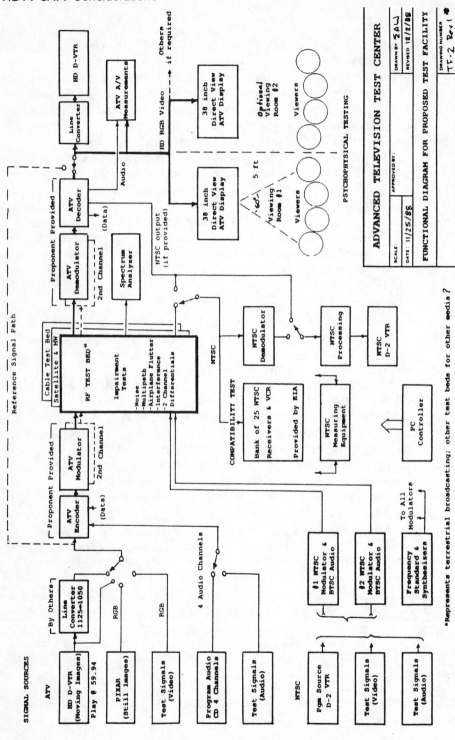

Fig. 7-1. A preliminary diagram of test sources, receivers, and signal paths illustrating Test Center HDTV operations (courtesy ATTC).

transmission formats. Citing consumer confusion between Beta and VHS, the Association warns that development of multiple, incompatible TV standards would retard ATV development and produce market instability. Members consider that HDTV receivers will be specifically affected by consumer friendliness of such sets. And while there is disagreement over any inter-operability of different transmission formats being accomplished via the marketplace, industry guidelines, or mandated regulation, some form of cost-effective interconnection is needed.

This is especially important between cable systems and the broadcasters. "Options currently under consideration," say the NCTA, are open architecture (all system) receivers or an external multiport interface connector.

SMART VERSUS MULTIPORT RECEIVERS

This subject and its attractions and detractions is likely to continue almost to the moment an HDTV system is selected. Some like the idea because it could accommodate virtually any system—and that's why an open architecture unit is called a "smart" receiver. Conversely, receivers with commonly-designed inputs might also handle a number of varied systems and not cost nearly as much as a totally computer-directed receiver. Thus far, it appears the academics are all in favor of the smart set and manufacturers prefer something simpler, especially because of cost.

The smart set could identify and receive various transmission formats using plug-in cards and special software programming for signal processing and reception. NCTA is concerned about significant drawbacks to any all-inclusive, open architecture approach to ATV. Cost would be high, in addition "an already expensive ATV receiver," could negatively affect consumer acceptance, and consumers may not wish to "sort out various receiver cards or physically install them."

On the other hand, a multiport connector or baseband circuit board could serve quite well, according to NCTA. However, for it to be effective, NCTA feels that a baseband component signal should be specified for all media. It should define the number of scan lines, field rate, and aspect ratio for all ATV transmissions. And this should offer an economical and practical means of connecting HDTV information from both feeder and distribution systems.

While the search continues for the best HDTV system(s), NCTA would encourage government investigation into methods of encouraging substantial participation from both American producers and manufacturers.

EIA INTERFACE

Still in draft form, but substantially ready for publication, the Electronic Industries Association has such an interface characterized "between NTSC television receiving devices and peripheral devices." Originally published as EIA Interim Standard IS-15, the Association is responding to cable television by "offering models" that will tune not only cable and broadcast channels, but also offer a baseband interface to accommodate a decoder designed for such an interface. This one is designed for baseband audio and video with a control bus. And since such signals are "relatively" standard, the same interface can be used with other peripheral devices such as video discs, VCRs, teletext decoders, DBS or MDS receivers, "and future consumer products." Used with a cable system, it will permit cable operators to reduce equipment costs to the subscriber and allow the subscriber full use of a TV having such an interface.

Such new receivers are to be designed with fast and slow automatic gain controls (AGC). In the fast mode, a 6 dB increase (RF increment) is to be equal to or less than 1 msec; while for a 6 dB decrease (RF decrement), the time is less than or equal to 2 msec. In the slow AGC mode, time constants are greater than or equal to 20 msec. Peak video transient change at pin 19, responding to two time constant changes are specified at less than 10%. The other 19 pin-ins and -outs are illustrated in Fig. 7-2. Titles are largely self-explanatory, except for several EIA highlights.

At pin 7, for instance, the dc level is not to exceed $\pm 1V$ into 75 ohms, and horizontal blanking must be present. At pin 8, the wideband audio output, the impedance is 600 ohms and frequency response at 3 dB down points equals 300 Hz to greater than 90 kHz. Amplitude ripple at 300 Hz over 50 Hz to 47 kHz should not exceed ± 0.35 dB, with phase response less than or equal to $\pm 3°$. Thermal is to be held to < -62 dB at 15 kHz (no video modulation) and buzz has to be less than -50 dB at 15 kHz. Total harmonic distortion has to be less than 2% for 70 kHz deviation at a 1 kHz rate (no video modulation).

The video format at pin 9 is used together with pin 20, which sets up the format. For composite video, pin 20 shows a 1, but for R/R-Y and B-Y, and Y/C, pin 20 is 0, or low. For the three inputs, in 9 the logic is 1, 0, 1. At pin 14, channel change and receiver power is indicated for a decoder. The decoder can sense signal dwell in the low state to either recognize channel change or power off.

Blanking logic levels at pin 16, the fast blanking and chroma input, are low for internal video from 0V to 1.2V and high (external) from 1.8 to 3V. The output will drive a 75-ohm load.

At pin 18, the sync tip is the most negative portion by less or equal to 200 mV, and the sync tip level is to be within 0.05V of sync level at pin 19. When authorized, the decoder shall restore a scrambled signal to sync tip output level that does not change between scrambled or clear operation. Decoder restored sync output is specified at 2.14V measured relative to 0 carrier level. At 100 IRE, video is to measure $2V \pm -.1V$.

The decoder response time's dynamic range can vary from 0.5 to 1.35V and the internal logic levels are supplied by T^2L, which needs a pull-up in the receiver only, thereby allowing a high state when off or open. Electrostatic discharge up to 1k is provided, and the logic has a fan out of three.

```
AUDIO       :                                    :
SELECT FUNCTION :        :  1                     :
                :                                 :
                :                       2  :  :  AUDIO INPUT
      AGC TIME  :                          :  (LEFT)
      CONSTANT  :        :  3                     :
                :                                 :
                :                       4  :  :  AUDIO GROUND
        SECOND  :                                 :
 AUDIO PROGRAM  :        :  5                     :
   SELECT LINE  :                                 :
                :                       6  :  :  AUDIO INPUT
                :                          :  (RIGHT)
          B-Y   :        :  7                     :
                :                                 :
                :                       3  :  :  AUDIO OUTPUT
                :                                 :
  VIDEO FORMAT  :        :  9                     :
                :                                 :
                :                      10  :  :  PERIPHERAL
                :                          :  COMMUNICATIONS
 LUMINANCE (Y)  :        : 11                     :
        INPUT   :                                 :
                :                      12  :  :  RESERVED
                :                                 :
    FB/CHROMA   :        : 13                     :
      GROUND    :                          :  CHANNEL CHANGE
                :                      14  :  :  AND POWER
                :                          :  INDICATOR
          R-Y   :        : 15                     :
                :                          :  FAST BLANKING/
                :                      16  :  :  CHROMA INPUT
                :                                 :
 VIDEO GROUND   :        : 17                     :
                :                          :  DECODER
                :                      18  :  :  PRESENT AND
                :                          :  DRS
RECEIVER VIDEO  :        : 19                     :
                :                                 :
                :                      20  :  :  PERIPHERAL
                :                          :  VIDEO
       SHIELD   :        :                        :
```

Fig. 7-2. Pin-outs and pin-ins of EIA's "draft" audio/video baseband interface for TV receivers (courtesy Electronics Industries Association).

In implementing pay-per-view operations, transmission speed is 1k bit/sec., modulation pulse width coding, with message lengths of 8 bits plus end-of-field. The first 4-bits in these transmissions are always 0001; the second 4-bits produce the data/command, and the end-of-field pulse is high for 3k microseconds—all of which requires a control 4-bit microprocessor. Binary, apparently, is reduced to hexadecimal code requiring only double digits or a single letter and digit.

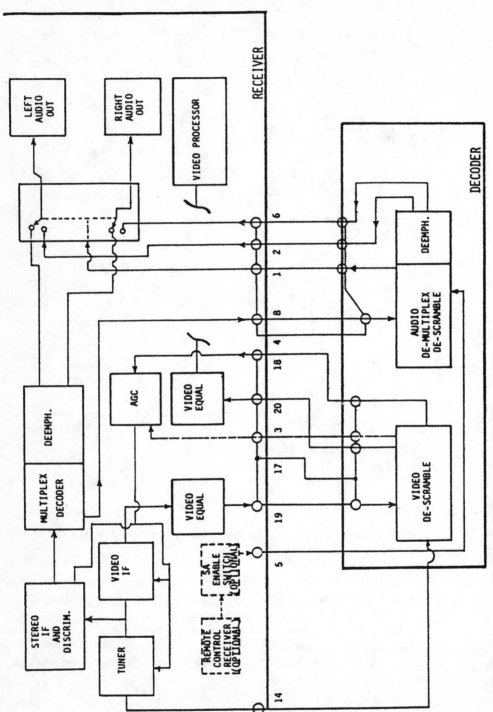

Fig. 7-3. Actual block diagram of the audio/video baseband TV interface (courtesy Electronic Industries Association).

A general outline of the receiver decoder is illustrated in Fig. 7-3. Presumably this is more or less the final format, and so you can expect to see an equivalent unit on your better (or best) receiver in the fairly near future. Will all this make any receiver qualified for HDTV? Not by a long shot, but it's certainly a good start toward standardized signal inputs from the various and sundry sources now operating. I suspect the satellite receivers will have to wrestle with D2-MAC, B-MAC and the two VideoCiphers® all by themselves.

CATV TESTING

While Cable Television Labs have not submitted a complete testing schedule as of the editing date for this manuscript, it is known that theirs will be a diversified setup for research and development, with contracts awarded to operating laboratories that are fully equipped to conduct specific tests and analysis. One of their first projects will be high definition television, and the second, fiberoptics. Thereafter, Cable Labs will continue to authorize further investigations of subjects and hardware of interest to CATV, backed by 80% of the cable industry and funded at approximately $7 million per year. Cable participants sign up for three years and will enjoy all the benefits of this consolidated search and discover operation.

8
HDTV Service Potentials and Anticipated Problems

MECHANICAL AND ELECTRONIC/ELECTRICAL DEVICES, WHATEVER THEIR NATURE and mission, eventually give trouble. What and how much certain difficulties could or should arise will probably depend on IC reliability as well as a thoroughly debugged design once it reaches market. Initial consumer products having radical changes often require at least a year's "settling in" time before genuine reliability can be honestly projected. And in the rush-to-market these new receivers with their special microprocessors and luma/chroma/sound decoding could very well present some interesting troubleshooting exercises before established procedures and familiarity triumph.

Considering that you're dealing with multichannel sound extensions of up to 6.5H, or $6.5 \times 15,734$ Hz (102.27 kHz) and video bandspreads up to 30 MHz or so, the two prime parameters have broadened considerably. Add multiplexing, 2- and 4-channel stereo, special sync and line-rate operations coupled with very large picture tubes or matrixed flat panels, and you have more than a few hard-nosed considerations to contemplate and surmount. When many of the very large scale integrated circuits (VLSI) are virtually imperial silicon mammoths, replete with tusks and very short tails, the going becomes somewhat more lumpy.

Further, you're going to suddenly discover that many of these hairy mastodons are completely enclosed in shielded compartments, where soldered commons (grounds, if you prefer) are highly important in preventing multiple transients from appearing in fast-switching logic. Under such circumstances, a few glitches here and there could become awfully misleading when looking for some cozy matrix or multiplex that just doesn't occur. Eventually, the manufacturers are going to have to generate a sort of system go, no-go checkout that offers the servicer some subsystem possibilities. But it is doubtful that the first of the breed will have arrived with such sophistication.

At least let's hope that the major ICs are stuffed in PC board mounts so they can be removed without the painful process of tricky desoldering and metallization ruin. Single-sided printed circuit boards are difficult enough, but double-sided boards, which are good probabilities, become double trouble when removing and replacing discrete and multipin components. True, in highly competitive consumer products, cost is an overwhelming consideration, but servicing must be given some thought too. Let's hope it is! At the same time, try and link quality with serviceability for a possible sampling of both worlds. With some well-directed perspicacity, perhaps you will. For those who won't or don't, they had best try another line of endeavor besides consumer products. The analog and digital mix becomes more integrated every day, and is no more than a forerunner of all-digital designs that are sure to follow.

When all this comes to pass, following a nice analog video or audio signal won't be a simple matter of oscilloscope or sound-detected signal tracing, but a 1s and 0s project that relates much more to subsystem analysis rather than discrete transistors or simple, 14-pin ICs. We haven't discussed these assumptions with American-operated TV engineering at any length so far, but such procedures have to emerge eventually as the only viable means of effective troubleshooting. In the meantime, baseband audio/video injected into digital inputs is one means of approaching this problem, depending on access to entry points and impedance-matching that could produce difficulties of its own. Both Zenith and Philips service departments are awaiting engineering material and directions that could begin to resolve such anticipated speculation.

In the troubleshooting portion of the chapter we'll see what can be done with the digital portion; and in analog, we'll illustrate a couple of subsystem approaches that should prove more than helpful—one will even show how to align video i-fs during on-signal, full-power conditions. This may not work in every instance, but it definitely will where your only "twiddle" inductor is connected to the synchronous video detector. As for chroma, a good clean-gated rainbow color bar generator and an inverting-trace oscilloscope will deliver an excellent chroma amplitude, phase and demodulator checkout for any color TV system. Note that we did not reference an NTSC generator—for this offers only various RGB levels at the cathode ray tube, rather than full chroma R-B displays in a readily decipherable vector wheel of 10 significant petals.

In the digital section of a hybrid (analog and digital) receiver, it may be possible to institute either or both logic and signature analysis of some description so that selected points can have identifiable bit streams to be used as references. Logic analyzers are single or multichannel test gear units with selected memory complements that arrange such bits into pages of selected lengths. Each page, then, can be displayed on a CRT for examination. A signature analyzer compresses these bit streams into a special display so that the overall sequence is or is not correct. In its most common format, signature analysis appears in modified hexadecimal, with digits from 0-9 and letters ACFHPU. Should such procedures come to pass, servicers will need one or more brand new pieces of test equipment, to learn and use for their survival. In the meantime, good, dual-channel inverting oscilloscopes, color bar and pattern generators, sweep generators, auto-ranging digital voltmeters, and 150 MHz signal generators might just come in handy. A few slumping and/or glitchy digits in a well-defined bit stream can wreak havoc with any analog-converted signal, be it voice or picture. Eventually the old ways will be missed, but new equipment and procedures could accelerate trouble determinations by factors of 10 (move the decimal point one place in an appropriate positive or negative direction). It all costs money, however, so put aside a few nickels (or kilobucks) for that fateful day when bitstreams and state or time domains are required techspeak in fully digitized consumer products. Fortunately or otherwise, that day has almost arrived. Are you ready?

OSCILLOSCOPES

Always my favorite topic, next to spectrum analyzers, there are many types and makes of these cathode ray tube display instruments with which you should have some familiarity, especially since *new* products are always appearing on the market with both plain and exotic claims and features. Oscilloscopes with vertical deflections from millivolts to 5 volts/div. are great for digital and low voltage analog systems, but they don't cut the mustard for systems with large deflection sweeps and high voltages. Preferably you should consider an instrument that has millivolts for the low end but between 10 and 20 volts/div. for the top end. Then, 8×10 or 8×20 V/div. and a $10\times$ low capacity probe can look at waveforms between 800 and 1600 volts, respectively. The ''8'' multiplier, of course, represents 8 vertical centimeters or divisions, however the graticule is calibrated. Horizontal-measuring graticule lines are always 10 in number, and a good set of figures would range between 1 or 2 seconds to 50 nanoseconds or less per division. Accuracies in both vertical and horizontal directions should not be less than 3%. This means that at least in the beginning, your scope should respond within 2% over most of its range.

Other scope considerations should include an adequately shielded *metal* package with carrying handle, plenty of integrated circuits for manufacturing repeatability and accuracy, ribbon (webbed) wiring for durability and low capacitance, and high quality printed circuit boards. The graticule should be lighted, high voltage should be highly regulated and specified around 15 kV $\pm$ 20%, and the sync portion should hold a rock-steady trace during *all phases* of video sweep, including the vertical blanking interval. A jittery waveform is about as useless as external mammary glands on a rattlesnake. The obvious admonition, therefore, is ''look before you leap'' into what appears to be deep water. There are often hidden rocks.

These aren't all the desirable features: a dual trace, triggered sweep scope is essential, there should be provisions for some sort of time base delay allowing special waveform examinations in expanded detail, and *both* vertical inputs should invert! And that's not idle chatter. For without this one feature you cannot make your scope into a vectorscope as we will most certainly do when examining the color circuits of either analog or digital TVs later on. Very few of today's instruments have this feature, so we'll select one that does, and does it well without bouncing its dc reference all over the countryside. This means that the dc level position should remain fixed from reference when the waveform is inverted. Otherwise you have a drifting dc level amplifier that can be annoying now and real problems later. The nice part of all this will suggest a reliable and full-featured instrument with 60 MHz bandpass for under $1,000!

THE NEW HM604

With vertical amplifier risetimes of 5.8 nanoseconds, and time base deflection in 23 calibrated steps to 50 nsec/div., plus a $10 \times$ magnifier to 5 nsec/div., there is even a component tester available on the front panel, a square wave calibrator switchable from 1 kHz to 1 MHz, and dual output voltages of 0.2 V to 2 V/div (Fig. 8-1). Line voltage is selectable from 110 to 240 V, and line frequency 50 or 60 Hz at a power consumption of only 43 watts. The cathode ray tube is a 15OCTB31 P43/123, with 12 kV accelerating, and graticule illumination (optional) in three levels. Measured weight comes to 17 lbs., including carrying handle which also doubles as a tilt stand. Vertical sensitivity begins at 5 mV fixed and variable to 2 mV, while bandpass for both vertical amplifiers amounts to 60 MHz at 3 dB down and X-amplifier for X-Y mode specified at 5 MHz. There is also an "active" TV sync separator for positive and negative-going line and frame capture in addition to the usual automatic triggering with filters for AC/DC/HF/LF and line positions.

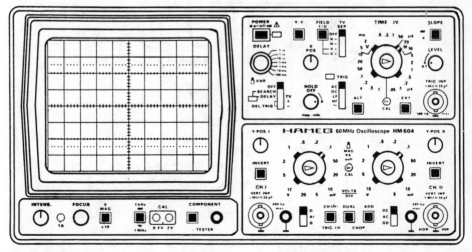

Fig. 8-1. Front panel of Hameg's HM604 60 MHz dual-trace oscilloscope with pseudo delay time base and 3% accuracy (courtesy Hameg, Inc.).

You will also find an after-delay trigger mode designed for highly stable asynchronous signals consisting of bursts, pulse trains, or unusual amplitude changes. Say you want more for your money? I seriously doubt if you can find it anywhere! Hameg has a good one here, and a HM604 is my personal scope used in all video and audio displays throughout the book where analog/digital and non-storage waveforms are required. You might also like to know that the latest in $10 \times$ low capacity probes have capacitor *and* inductor adjustments for both low and high frequencies. This prevents undesirable rolloffs in Hz and over/undershoots in MHz. With the onset of digital rectangular waveshapes and very sharp risetimes, any helpful, accurate compensation should be welcome.

It's interesting to note that precise high frequency compensation is only possible with square wave generators whose risetimes are less than 5 nanoseconds. So you do have to be careful when working among digital circuits that your own equipment is not at fault. Remember also that oscilloscope displays are always in terms of peak (P) or peak-to-peak (P-P) (Fig. 8-2), depending on whether the voltage swings positive or negative from some common or dc reference, or whether it swings about some center potential, usually dc or 0. To reduce these peak displays to root mean square (rms) measurements, the first is divided by 1.414 and the second by 2.828. You will now have the effective value comparable to that developed by direct current. For instance, were you to float your scope across the ac power lines (the British call them ''mains''), you should see approximately 340 volts. Divide this by 2.828 and the answer is 120.2 volts rms. Most useful measurements, therefore, are generally described in terms of rms so they can be referenced in either ac or dc applications. But don't confuse sine waves with pulses since they are different breeds.

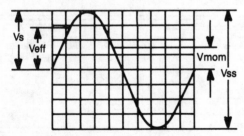

Voltage values of a sine curve

V_{rms} = effective value; V_p = simple peak or crest value;

V_{pc} = peak-to-peak value; V_{mom} = momentary value.

Fig. 8-2. Sine waves and their various operating measurements. RMS = V_{eff} (courtesy Hameg. Inc.).

A pulse (Fig. 8-3) delivers voltage or power in terms of its width, amplitude, and the time of a single cycle:

$$E_{rms} = E_{max} \sqrt{t_d/T}$$

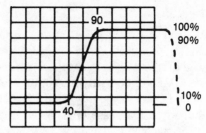

Fig. 8-3. Rise and fall times of any rectangular wave or pulse are always evaluated at the 10-90% points (courtesy Hameg. Inc.).

This means that the square root of the t_d time duration of one pulse divided by the T time of one cycle multiplied by the maximum voltage of that pulse equals the ultimate rms value. The power of such a pulse, then, could become: P = E^2/R (or I^2R, or EXI), with R being whatever resistance E^2 has developed across. Obviously the answer is always less (in voltage) and E_{max} since t_d and T represent some fraction less than 1.

It may also be useful to know that the duty cycle of any pulse is equal to its duration times its repetition rate: $t_d \times f$. And during these exercises recall that Time = 1/Frequency and F = 1/T. In other words, they are the exact inverse of one another. For instance, a vertical repetition rate of 59.94 Hz translated into time becomes F = 1/T, or T = 1/F, and 1/59.94 Hz = 16.68 milliseconds, or the time of a single TV field. Double the field rate and you have a frame, of which there are 30 in each second. Similarly, a full horizontal line sweeps in 63.5 μsec. Therefore,

$$T = 1/F, \text{ and } F = 1/T = 1/63.5 \times 10^{-6} = 15,734 \text{ Hz}$$

So you see, these "incidental" items all tie together in the video world; and if you can't measure them adequately and accurately, there's no way to tell poison from precision. An uncalibrated oscilloscope, for instance, is pretty terrible in analog, but a complete disaster in digital; so be forewarned and check both vertical and horizontal calibration with at least your built-in amplitude calibrator and time base with a good electronic counter. Our Hameg also has both ramp and video outputs, so any suitable vertical scope input should track time base calibration accurately with counter connected. If your scope hasn't an output, then a Y voltage divider at the scope's input and a counter will permit adequate frequency accuracy determination, and you can tweak your scope accordingly. In older units, a compromise between low and high time base settings is almost inevitable. And if time/frequency tracking is too far off, either have your equipment repaired, or acquire a new model. An antiquated oscilloscope belongs in the same class as an old vacuum tube voltmeter; they're both trash!

Suggestions and Cautions

I would like to warn you that when pulse and rectangular (or square) wave information requires investigation, only about a negative factor of 10 is possible. As Hameg cautions, due to the harmonic content of these signals, a maximum repetition rate of some 6 MHz in 60 MHz is accurately possible. So if you're doing 20 MHz logic repetition rates, you may need a 200 MHz oscilloscope for complete accuracy. Burst signals are also dogs in the trigger/stability department. Variable time division, hold off, and even various filter selections are probable, depending on amplitudes, time, and durations—our old pulse buddies. Here, as long as low frequency rolloff doesn't occur, use dc amplifiers if possible, unless there are peculiar dc levels or special biases. At high frequencies, capacitors become short circuits, but inductors induce preshoots and overshoots with a vengeance, often causing false triggering if their pulse widths aren't sufficiently narrow. In HDTV you can count on some pretty rapid MHz clock and switching among special decoders in the video portion. This is another well-supported reason why some sort of signature (condensed) logic checkout needs to be developed during the 1990s. Gigahertz oscilloscopes are not inexpensive, but then again logic and digital oscilloscopes should do the trick. Just now there are none of these new HDTV receivers to investigate. But our digital receivers should offer some initial preview, and we will certainly illustrate what's immediately available, including the relatively slow logic that accompanies them.

In our familiar analog scope domain, dc plus peak voltages should ordinarily not exceed 400 volts for any length of time, especially at higher repetition rates. If such limits are exceeded, the low frequency compensating capacitor across one series input resistor may turn red and fade away, damaging input circuits of the oscilloscope. And where there are inductive transients in addition to higher frequencies, even a circuit touch or two around high voltage can ruin a probe very quickly, losing that special 10 MHz isolation and its desirable protective and range-adding impedance. If the ac/dc level is constant, however, you can go considerably higher than 400 volts with relative safety. But don't press your luck, some equipment is just plain fickle. Should systems require 1,000 and 2,000 V measurements, then purchase a 100:1 probe that's suitable for your particular instrument. You should also be reminded that good commons (grounds) are essential to any and all measurements. In TV receivers, high voltage pickup is always possible if the coaxial shield portion of the probe isn't securely fastened to system ground. And any time you do spy a large glitch in either a spectrum analyzer or oscilloscope display, measure its frequency immediately. Chances are this is a 15,734 Hz spike of voltage that's found its way through your input cabling. Once again, secure the ground or use better shielded cable.

The foregoing is basically all you want or need regarding our friendly oscilloscope. It's great for detected information and radio frequency measurements in the low megahertz. But into higher RF regions you'll need either a sampling oscilloscope or spectrum analyzer—the latter being much preferred since it is directly calibrated in terms of frequency and has great vertical sensitivity. But most readouts are in decibels (dB), which are tenths of bels and primarily reflect measurements in voltage, power, and frequency. The oscilloscope, too, can have its readouts calculated in dB, whether involving current, voltage (E/I), or power (P). Here's how...

For E or I: dB = 20 log E_2/E_1 (or I_2/I_1)
For P: dB = 10 log P_2/P_1

As you can see, there must be two of everything, one *in* and one *out*, to supply a ratio. Then you have a comparison between the two and can take the logarithm of the difference on any worthwhile calculator, multiplying the result by its proper factor. This is especially useful for stereo audio and channel separation measurements. Should the input be greater than the output, a negative ($-$) sign is placed before the 10 log or 20 log portion, and the answer is -41, -10, or whatever. So instead of having a gain, you have a loss between input and output of some finite figure measured in decibels. You'll find it very handy. While radio-audio may have considerable separation dimensions, TV multichannel sound rarely exceeds 25 or 30 dB. And in some instances, sorry to say, the log difference is hardly 12 dB. Remember, however, that both oscilloscope channels *must* be calibrated equally for accurate measurements. Guestimates are not even poor excuses—in this business they just don't work.

SPECTRUM ANALYZERS

Once you've made a friendly acquaintance, spectrum analyzers aren't nearly as formidable as they seem. Actually, the newer ones with superb microprocessor controls, menus, and dc-guarded inputs can produce useful data their forerunners never approached even in visionary concept. How would you like to look at video pictures, hear audio, dial in carrier-to-noise to four places, place visible markers wherever, and look at satellite transponders from 22.3 kilomiles above the equator up to 20 GHz with only good reflectors and reasonable amplification? If all this sounds good, bear with us, you'll be working with frequencies from dc to 1.5 GHz and many points in between. Once the identifying terminology and general concept sinks in, applications are reasonably straightforward for standard measurements. When you've passed that point, however, menus and subroutines do become a bit difficult especially in complex measurements where only a good operator's manual will save you.

Primarily available in three definitive classifications, spectrum analyzers operate swept-tuned, realtime, and fast Fourier transforms. Respectively, they become tuned-filter heterodyne RF receivers, filter-bandwidth frequency selectors, and Fourier-computing transformations. Instrument costs (for the decent ones) range between approximately $9,000 and $100,000, depending on applications, accuracy, resolution, functions, sensitivity, auto calibration, dynamic range, and programmability. The more exotic the characteristics, the more you pay. An analyzer with excellent specifications operating between 10 kHz and 21 GHz without external adapters markets at over $40,000.

Heterodynes, the wideband, realtime units, are frequency limited with special bandpass filters, and the Fourier transforms often are restricted to kHz frequencies, especially useful in frequency domain measurements. So we'll stick with the broad-frequency radio front end instruments that have been so helpful between kHz to GHz. The particular one primarily worked with is an 8th wonder all by itself.

TEKTRONIX 2710

Like its brother/sister family brood, the 2710 (Fig. 8-4) measures signal strength in terms of dBm, dBmV, dBμV, etc., as a function of frequency—up to 1.8 GHz—and at bandwidths/division of 180 MHz or less. Resolution bandwidths are adjustable from 3 kHz to 5 MHz in indicated steps as they supply inputs to audio and CRT display sections that follow. In the power-up self test, LED lamps turn on, and you wait during standby as the analyzer completes its task. If parameters aren't within specs, you're very likely to see *normalization*

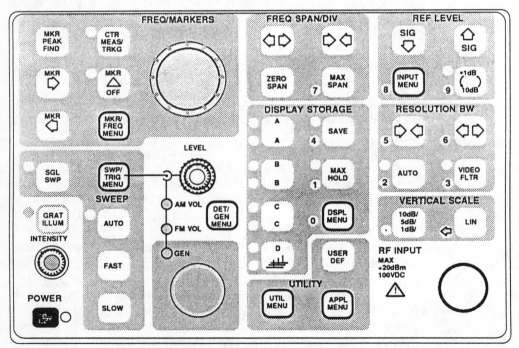

Fig. 8-4. Front panel drawing of Tektronix' Model 2710 microprocessor-controlled, low-cost, spectrum analyzer useful for both video and satellite signals with passband from 10 kHz to 1.89 GHz (courtesy Tektronix, Inc.).

suggested. You promptly respond by pushing the proper buttons and normalization occurs.

But while you can display a spectrum with only three controls, the various menus allow such things as special parameter setups, markers on and off, input signal levels, triggering, scale factors, resolution bandwidths, video filters, max or zero spans, storage, and utility applications. If these aren't enough, submenus and extra options are available with little more than the touch of another button for specialty operations within selected menus. Basic measurements are easy, but subroutines for frequency, normalization, carrier-to-noise (C/N) for both AM and FM modulated carriers, measuring parameters, and so forth do require some thought and specific commands. But if initial settings aren't satisfactory for some measurements, factory-stored settings are always available on startup and elsewhere, as illustrated in Fig. 8-5. Afterwards, you can customize whatever you like within a very considerable range of options. For instance, how about ensemble averaging at max, mean, or min, number of averages, saves, and terminations? And that's just the beginning. Results can be displayed from registers A, B, or C. In the video display portion, you can trigger on line, field, or single sweep, detect broadcast or satellite pictures, and select sync/video polarities.

Span/Div	=	MAX SPAN
Center Frequency	=	900 MHz
Center Frequency Corrections	=	ON
Signal Tracking	=	OFF
Resolution Bandwidth	=	AUTO (5 MHz)
Sweep Time	=	AUTO (10 msec/div)
Tune Increment	=	AUTO
Triggering	=	FREE RUN
Reference level	=	+20
Reference Level Units	=	dBm
Input Impedance	=	50 ohms
Preamplifer	=	OFF
Minimum RF Attenuation	=	0 dB
First Mixer input Level	=	−30 dBm
RF Attenuation	=	AUTO (50 dB)
External Attenuation	=	OFF (0)
Vertical Display Mode	=	log
Vertical Scale	=	10dB/div
5 MHz Video Filter	=	OFF
Display Register	=	D
Max Hold	=	OFF
Max/Min mode	=	ON
Video Sync	=	Positive
Video Polarity	=	Negative
Markers	=	OFF
Graticule Illumination	=	OFF

Fig. 8-5. Factory analyzer settings as they appear on power-up unless overridden by operator stored commands (courtesy Tektronix, Inc.).

At the moment, you won't investigate the various measurements and their readouts, the "live" demonstration portion of the chapter will amply illustrate what this analyzer can do within its frequency range. In audio measurements, we'll use the old, tried and true 7L5, whose frequency span goes from dc to 5 MHz. Between the two, and possibly a readout or two from my 7L12, you can cover all frequencies between dc and 1.8 GHz very adequately. See if you don't agree.

However, to visualize the considerable flexibility and abilities of the 2710, a descriptive block diagram of the analyzer might prove informative (Fig. 8-6).

Supplied by Tektronix, an RF input up to 20 dBm and 100 V dc max. is permitted, followed by an RF attenuator in 2 dB steps from 0 to 50 dB. A preamplifier of some 20 dB can be switched in, if required, followed by the 1st converter and its YIG oscillator and frequency control. Then comes the 2nd and 3rd converters, 1 dB and 10 dB step selections, and the various resolution bandwidth possibilities, with auto or selective sweep trigger control. Video and audio detectors follow RBW stages with either baseband outputs or continuation through scale factors, storage and display. At the bottom of the diagram you see a block for the front panel keyboard, the microprocessor and the GPIC, RS232 bus interfaces. The input calibrator supplies excitations at −30 dBm and 100 MHz. Mentally, couple Figs. 8-4 and 8-6, and you have a pretty good idea of the operational abilities of this outstanding analyzer.

SWEEP/SIGNAL GENERATORS

"Something old, something new, something borrowed . . . " and more signals to stew. Not the best of rhymes, but with practical intent, nonetheless. This new breed of television receivers will, more than ever, require wider i-f passbands, more precision tuners, and several other requirements probably not yet designed. This means a formerly-valued piece of venerable test gear will have to be dusted off and calibrated, new signal and pattern generators procured, one or two old books resurrected on sweep and signal troubleshooting and alignments, and a whole flock of analog (in the beginning) and, later, all-digital generation and detection methods with which to deal. Only wish I could cover them all, but uncertainties make that totally impossible.

We can, however, work slightly backwards, offering insights into what used to be, then go forward with a few current examples, and conclude with several possibly "educated" guesses based on some of the problems discovered by the Advanced Television Test Center (ATTC).

The Sweep Generator

Recalling that a spectrum analyzer is a tuned RF and sweep generator, calibrated vertically in dB and horizontally in frequency, why not use a sweep generator in one or two limited applications as the horizontal exciter portion and

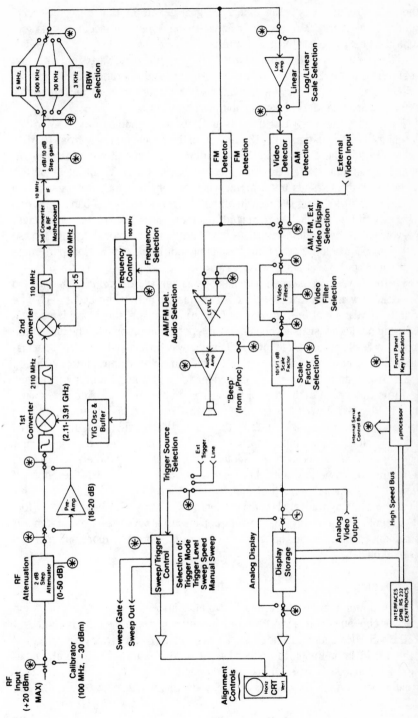

Fig. 8-6. Signal flow block diagram of the 2710 and its main operating divisions (courtesy Tektronix, Inc.).

171

either available markers or an external signal generator that's carefully calibrated for the marker? A relatively linear sweep input and accurate markers might just give you everything you want.

Using both an old Sencore and a B & K-Precision sine wave generator (on the shelf), with the Sencore recently resurrected and operating, see what can be done after an hour's warmup and an electronic counter check tacked on to a linearity examination for flat sweep and marker accuracy.

Fortunately, this particular unit has an RF output variable from 10 μV to 100 mV and guaranteed flatness of $\pm$ 3 dB over six U/V bands ranging from 10 MHz to 920 MHz (on the 4th harmonic). In addition, there are an even dozen markers and added input for an external marker that has 24 mV output or better. The sweep width extends to 15 MHz, which is very adequate for present and future purposes if TV signals remain within the NTSC-assigned bandwidth of 6 MHz, even if two channels are ganged together. In an instrument of this age, however, you're bound to accumulate some drifting and other deterioration, so don't demand absolute perfection. It's simply not there. But for ballpark measurement purposes, let's see what it can do.

Fortunately, we're blessed with a scope having an *external video output*, so the markers won't be loaded by the electronic counter when calibrated. The scope's 10 MHz input impedance is sufficient to prevent shunt problems from obscuring markers visible at the generator's port. But unfortunately, unless the counter is designed to handle low level analog waveforms and high frequencies, you're SOL. Therefore you have to work with an external input into the sweep generator and an inexpensive signal divider for electronic counter monitoring due to non-precision calibration of the signal source. A good example is illustrated in Fig. 8-7 showing the three 44.25, 45.75 and 47.25 MHz markers with a large check marker positioned on the right. When calibrated, the external marker is simply superimposed on each of the other three, one at a time, ensuring an accurate frequency readout on the electronic counter. Yes, this all takes time and some ingenuity, but results are completely worth the effort. If the sweeper's voltage regulation holds within at least millivolts, this careful recalibration could figuratively last many months or even years, believe me! I forgot to mention, by the way, that power supplies are always measured first in *any* instrument calibration and repair. This will become apparent more and more as you read on.

Refurbishing

With power sources fully qualified and adjusted, the marker calibration endeavor continues with acceptable results until you reach the 6 MHz (Fig. 8-8) and the 4.5 MHz sources. These are a pair of 2N5172 multivibrators triggered by the 1.5 MHz multivibrator, which, in turn, derives its drive from a crystal-controlled 500 kHz source. Ordinarily, you might think that this arrangement is virtually foolproof. You should be so lucky! In this instance, we weren't. The

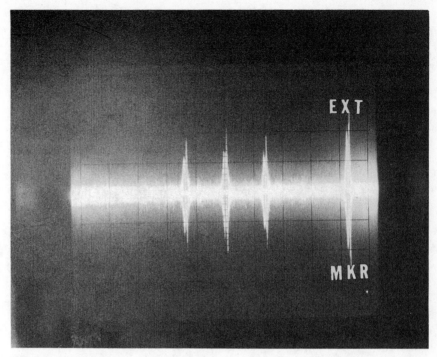

Fig. 8-7. A trio of crystal-controlled markers at 44.25, 45.75, and 47.25 MHz.

4.5 MHz oscillator calibrated reasonably well—at least to four places, but the 6 MHz one would only read 5.732 MHz.

Before you go tearing into the 6 MHz circuit, however, remember that from the 1.5 MHz fundamental, this vibrator must produce a usable harmonic to supply the 39.75 MHz upper adjacent picture carrier trap marker, while the 4.5 MHz delivers an initial 3rd harmonic output for the 41.25 MHz sound carrier. Furthermore, both of these markers have very small ultimate higher frequency amplitudes and are difficult to see on baseline sweep. Actually, they're only half the amplitude of the three illustrated in Fig. 8-7. Do you have a problem? You betcha! Let's see how we go about a solution.

You could have a 1.5 MHz clock drive problem or the power supply and/or load resistors could be increased in value. A scope with dc amplifiers or a good digital voltmeter will tell the initial story, and then you can proceed from there, bearing in mind there may be loading of both these multivibrators from some output source, their signals must mix with the main 45.75 MHz crystal-controlled origination to deliver the final difference frequencies of 39.75 and 41.25 MHz. These ultimately reach a beat-bridie amplifier for selectable output, controlled by a potentiometer for marker height and an LC tuned arrangement for marker width. So a paper analysis isn't really all that simple without a few substantiating measurements. Just remember that other markers in the series are

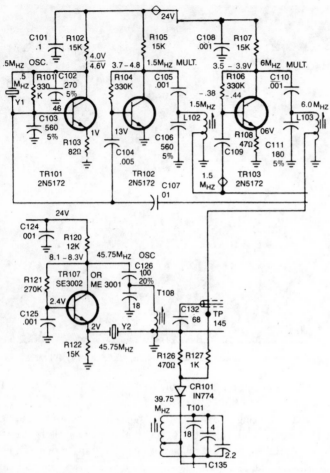

Fig. 8-8. An example of a sweep generator combining dual crystal-controlled oscillators and their accurate product markers.

all on-frequency and have plenty of amplitude. This is exactly why troubleshooting is characterized as both an art and a science. To be good you have to know your electronics and be artful, too! Any accidents or progeny along the way are your concern.

Nonetheless, there is a small clue that you might easily overlook. Since you're working with the lower marker frequencies that are in trouble, observe that the greatest mixing with the 45.75 MHz primary source takes place in these two combining (subtracting) networks and any unusual load here could well be the problem. Let's see.

On a pure hunch, we reconnected the electronic counter to the scope's video output and again measured the frequency output of both the 4.5 and 6

MHz multivibrators. By tuning the 4.5 MHz core just a little, virtually full amplitude returned to its output and the 6 MHz multivibrator could be tuned to 6 MHz. However, the amplitude of this higher frequency measured only in the low millivolts, rather than about a volt, while the 4.5 MHz output not only was on exact frequency but produced an amplitude of over a volt!

Now you know that the multivibrators interact in their tuning, dc measurements appear approximately correct, but the drive output of the 6 MHz unit was at least a factor of 5 low. Is this due to a slumping transistor or is the combining network at fault? It won't take long to find out. But it might be a good idea to replace TR103 anyway, since undue loading causes heat, and a hot transistor flows more than its share of current. However, being a small signal transistor, a finger on the casing won't prove a thing. In this instance, replace, making sure the new one has plenty of gain. You are undoubtedly aware we could make a preliminary check with an in-circuit transistor checker, but because this is an oscillator with inductance, resistance, and capacitance all in the network, most readings will not supply worthwhile accuracy. But you may want to check the new transistor's *gain* before placement. A Beta gain of about 200 would be just dandy.

TR103, fortunately, is an old and rather common transistor that is readily available at Radio Shack under the Archer cross reference of 279-2009. And yes, it does have a gain of 200. This, therefore, combined with input clock capacitor C109 did the trick. With greater difference subtractions from the 45.75 MHz main oscillator and the 6 MHz generated through harmonics, your signal output will be smaller, and only careful tuning of tapped inductor L103 the 39.75 MHz transformer below does the rest. All eight markers from 39.75 to 47.25 MHz illustrated in Fig. 8-9. Our "X" problem child is on the far right, sufficiently out of noise, but also smaller than the rest. The remaining markers are pretty much the same amplitude and will do very well on the response curves. Tuning all these is positively a chore, but it must be done for the sake of accuracy. Otherwise, your final tuner/i-f response curves will certainly be inaccurate—and that won't do.

Markers such as 39.75 MHz, 41.25 MHz and 47.25 MHz are on or near the baseline of the swept response curve and easily seen. Others on the slopes should be turned 90° for positive recognition and placement. Otherwise, there's every probability of a high amplitude, wide marker located a few MHz out of position—a catastrophe that can narrow video bandpass, wipe out color, depress sync, and/or move audio into luminance.

CHECKOUT

A simple sweep demonstration is now in order with our carefully calibrated generator—all without bias injection since we have neither tuner schematics nor i-f test points. You may think this is a relatively poor way to go, but it will certainly confirm whether the tuner is or is not working, and on what selected channels. Thereafter, we can sample the i-f output and see what that brings.

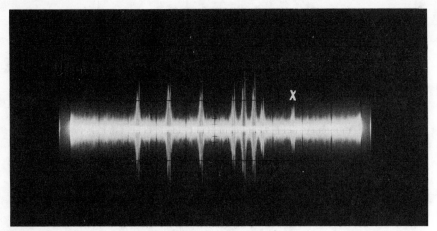

Fig. 8-9. All eight markers now appear at their appointed places. The 39.75 MHz marker is identified by the small, white "x."

The receiver was first isolated from its power source by a transformer-tuned variac, then *channel 4 was selected*, the sweep generator connected to the receiver's RF terminals through coax and then to our Hameg oscilloscope through video and horizontal (sweep) inputs.

The tuner/i-f cable was next disconnected from the i-f and a swept input detected via the generator's detector input. The 67.25 MHz video carrier appeared almost exactly in the center of the waveform, which had little symmetry (Fig. 8-10). By careful adjustment, the sweep width allowed a semblance of the humped response curve to appear, but with the video carrier still at approximate center. Using an external generator, we added the 71.75 MHz audio carrier at peak on the left, so that a fairly passable tuner response was generated. Its amplitude measured exactly 4 volts at highest peak. Note that markers distort the curve not at all.

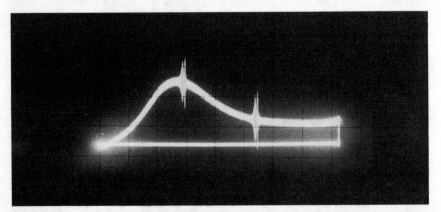

Fig. 8-10. Recognizable but imperfect tuner swept response does show the tuner is operating.

In this way you can check any channel for proper operation with adequate markers (if available), and also obtain a general idea of waveform symmetry. With proper bias information, I'm sure a virtually perfect response could be produced. Considering this is one of Zenith's brand new 1989-1990 digital receivers, you can be sure the tuner has ample bandpass for almost anything. Next, we'll work on the i-fs to see what can be done there.

The tuner cable is now disconnected and the sweep generator input connected directly to it. With a bit of hunting we found a detected video output (by first monitoring off-the-air video signals), and connected the sweep generator's input to this test point.

By using an external marker (center of response) we found the bandwidth was only 2 MHz wide, markers would not fall in place, and the lowest marker possible on the curve turned out to be 45 MHz, regardless of trace amplitude, sweep width settings, or anything else. As Fig. 8-11 indicates, there is already a partial overload from too much amplitude. The 47 MHz marker isn't on the baseline, and the midpoint marker is only 46 MHz. So this receiver won't respond without a little dc for its ac curves. Let's hope the next one you tackle will have all the information attached.

Still without schematics, but bolstered by frustrated determination, we did find a test point on the i-f board marked AGC. With a 10k current limiting resistor in series with a regulated dc power supply, and careful sweep generator tuning, the curve in Fig. 8-12 was finally produced. It's not the greatest, but it will have to do without bonafide factory instructions. At least there were no disconnects or other gyrations, other than feeding a 45 MHz signal with markers

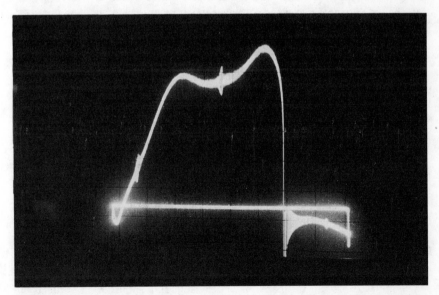

Fig. 8-11. Faked i-f curve has a bandpass of only 2 MHz that begins at 45 MHz and ends at 47 MHz. Beware of all these!

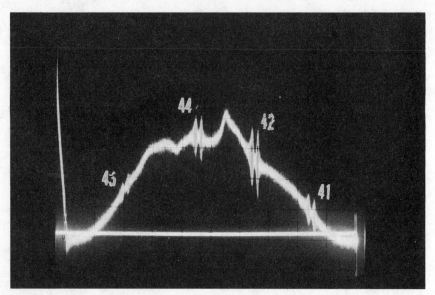

Fig. 8-12. Without manufacturer's instructions but with 4V AGC bias, an i-f response is possible with fairly adequate markers but sloppily shaped.

directly into the i-fs. Furthermore, the markers did, indeed, fall into place as you can see on the curve, with both 45.75 luminance and 42.17 chroma frequencies at approximately 50% points on either side of the response. True, this doesn't absolutely prove that all's well within the i-f under all circumstances, but the i-f is positively working within swept and marker limits. So the i-f spectrum is definitely under control. The point is pretty well proven that even digital receivers can be spot checked for spectrum display with a little imagination and considerable care. The old sweeper and an external B & K-Precision Model 2005 RF sine wave generator did the rest. The 42 MHz marker, by the way, is that of the 2005. Very clean and very handy as you can see.

OTHER SIGNAL GENERATION

Now you're in for a little surprise! Tuner and i-f matches and i-f alignments can be checked with *multiburst*. What's multiburst? It's a series of separate oscillations extending between 0.5 MHz to 4.2 MHz, and usually included on the better NTSC generators, especially B & K-Precision's Model 1260. Any worthwhile television set should deliver reasonably linear multiburst through its video detector, but the best allow these oscillations to be readily seen at the cathodes of the picture tube. Cheap sets with chroma traps in luminance, cut off resolution and definition at 3 MHz or less to prevent chroma-luma interaction and crosstalk. The better sets have CCD (charge coupled device) or glass delay line comb filters for such separation and, consequently, don't reduce luma bandpass by 25-50%. In all receivers, however, inserting multiburst into the antenna terminals on an appropriate channel can indicate satisfactory alignment or needed realignment due to nonlinearity or frequency cutoff.

An example is illustrated in Fig. 8-13, with this same Zenith digital receiver. Now, its luma bandwidth is "no secret anymore." As you see, a full 4.2 MHz bandpass shows at both the video detector (bottom trace) and at the CRT's red cathode. A falloff at 4 MHz is usually conventional for even the best receivers and amounts here to a 6 dB loss in amplitude. However, a full 4 MHz bandwidth is assured by the remaining burst. When aligning, make start-to-finish multiburst oscillations as uniform as possible.

Regardless of multiburst's utility, color bars in their original form seemingly disappear at the cathode ray tube even though they are apparent on its faceplate (Fig. 8-14). This is due to excitation levels required for the various colors rather than a simple analog trace. Remember also that chroma and luminance have been separated and then reprocessed prior to CRT cathode entry.

To make color information completely useful, we'll need a clean, gated-rainbow generator with 10 spokes or "petals" for any meaningful color analysis and troubleshooting. We therefore resurrect another Sencore product from many years past to do the job. And because there is no added gating between the color excitations, the red and blue vectors—not green because it isn't at quadrature with the others—form a color wheel from which to draw a number of substantial conclusions. Still working with the Zenith digital set, here's what you see at the first video amplifier and then at the blue cathode of the picture tube (Fig. 8-15). You can't do much with this one except establish that you have 11 color excitations at the video detector output and 10 "hounds teeth" since

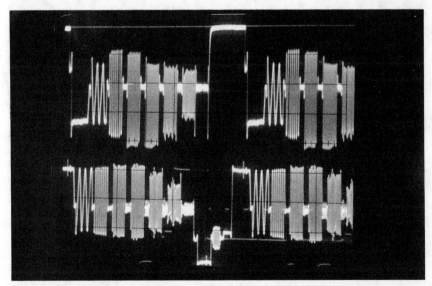

Fig. 8-13. Multiburst from 500 kHz to 4.2 MHz immediately reveals luminance passband between video detector and CRT input.

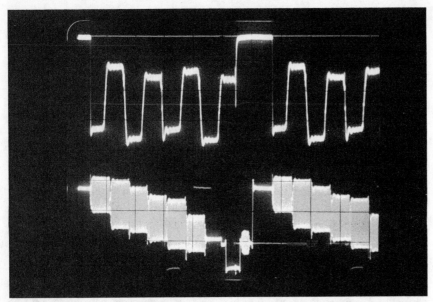

Fig. 8-14. NTSC color bars are essentially perfect at the video detector, but only a series of levels at the CRT.

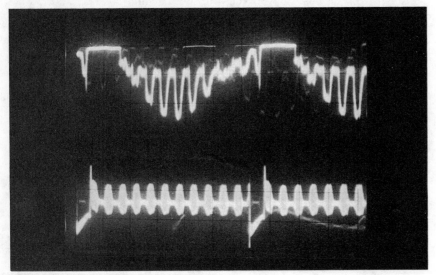

Fig. 8-15. A blue-gated rainbow signal at the top and 11 color bars at the video detector below.

one color is eliminated due to blanking. Note also that nulls occur in the blue waveform at the 3rd and 9th bars.

The information in Fig. 8-16 is the one we've been waiting for. The 3rd and 9th bar nulls appear again in blue, while red has only a single null at bar No. 6. In

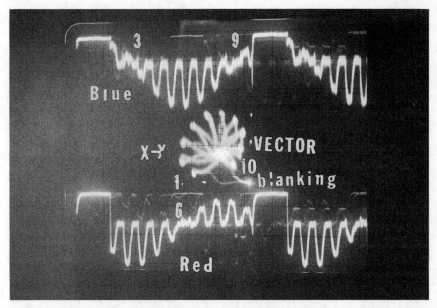

Fig. 8-16. Blue (top) and red (bottom) chroma demodulator outputs into the CRT provide highly useful information, especially with the resultant vector in between.

the center is your vector that produces considerable information. Note that all petals are approximately 30° apart and number a total of 10. There are no crossovers among the voltage excursions but they do differ in widths and numbers 9 and 10 are shorter than the rest. This indicates that in detection, linearity isn't exactly perfect, and there may be just a little problem with the green matrix amplifier not quite doing its job. Had all petals been as broad as those from 7 to 10, the chroma bandpass amplifier would be faulty; and in the older, manually-tuned receivers, an exceptionally broad-banded vector meant channel mistuning.

As you can see, the position of this vector is still phase-shifted 90° from normal since the X-amplifier will invert, but that doesn't help another quadrature turn so that blanking falls in place of petal No. 1, and the entire vector is rotated 90°. (We thought this problem had been licked with dual Y amplifier inversions—guess not.) Each of the 10 petals represents a color, and here's what they are, beginning with petal No. 1: yellow orange, orange, red, magenta, reddish blue, blue, greenish blue, cyan, bluish green, and green. Cheap receivers squeeze oranges and reds together to broaden fleshtones while moving blue (no. 6) from its normal quadrature position with respect to red to between 80° and 85°. The 11th and 12th petals, incidentally, are lost to blanking that occurs between green at 300° and yellow orange at 30°. If half the vector is missing, you've lost either the red or blue final amplifier. When both fail, there's no color and probably no picture since luminance and chroma recombination usually occurs in the final driver-amplifiers before the cathode ray tube.

A worthwhile low cost signal generator with fundamentals to 150 MHz, good stability, decent output, and clean sine waves has become a valued acquisition for video/HDTV/satellite analysis. New from B & K-Precision, the 2005 is an AM signal source with 1 kHz internal modulation but will accept external modulation between 50 Hz and 20 kHz. It has both a variable RF output and separate frequency monitor output, the latter leveled at 50 mV rms from 100 kHz to 150 MHz in six switchable frequency bands. RF begins at 100 kHz and extends to 450 MHz on third harmonics, although outputs in this final harmonic band are weak in amplitude. AM alignment within the frequency bands presents no problem, but this generator will also do FM alignments if the cw carrier is tuned to 10.7 MHz and monitored for accuracy with an electronic counter. Minimum RF is suggested. Unfortunately, we did not have a schematic delivered with this instrument, so a photo will have to do (see Fig. 8-17). It was absolutely invaluable in our alignment procedures, especially for checking markers and response curves. It should also become highly useful in verifying video bandpasses for both terrestrial and satellite receivers, as well as RF for short wave sets. At an attractive price, this little 5.5 lb. unit is both accurate ($\pm$ 3%) and handy.

The B & K-Precision 1260, of course, is a premier color bar and pattern generator having all sorts of outputs and inputs, and can accurately test almost

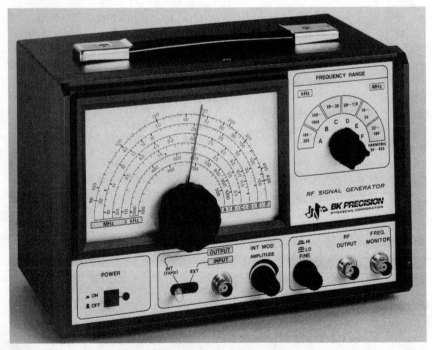

Fig. 8-17. Our "indispensable" 150 MHz B & K-Precision 2005 signal generator so useful in sweep marker calibration and general RF measurements (courtesy B & K-Precision).

any color equipment with NTSC specifications and bandpass. The multiburst portion, however, has a variable extension from 4 to 8 MHz and will easily identify bandspreads found in jackpacks and monitors. The 2005 and the 1260 will probably become a useful combination to explore wideband HDTV once equipment is in production, the latter having *both* external audio and video inputs in addition to RF on channels 3 and 4, as well as dothatch, vertical and horizontal bars, and red, blue, and green fields for color purity and (combined) CRT purity checks. Rather expensive but necessary for good color analysis and troubleshooting.

MEASUREMENTS

Having begun with a few instrument descriptions and applications, let's now begin seriously to apply a number of meaningful tests and applications to RF and audio, including satellite signals, with which you'll be contending as new and better equipment reaches the market. Instruments used will be a B & K-precision Model 2009 MTS TV-stereo generator, Hameg's HM604 oscilloscope, a Mod. 2710 Tek. analyzer for RF, and a Tektronix 7L5 audio-video analyzer inserted in a 7613 mainframe. The reason we're specifying these instruments is so you can duplicate apparent results by using the same or equal test gear under similar circumstances. We also prefer to be authentic and specific rather than deal with academic alchemy.

RF FROM THE SKY

Radio frequency waves have been in man-made existence since 1888 when first demonstrated by Heinrich Hertz, and this is why we call their excitations ground, sky, troposphere, and space waves "Hertz," instead of cycles, in his honor. Ground waves are vertically polarized, whereas sky and space waves have horizontal, vertical or circular polarizations, and their common frequencies range between 10 kHz (10×10^3) to more than 30 GHz (30×10^9). Television broadcast frequencies extend from 54 MHz to 806 MHz (channels 2 through 69), while commercial Fixed Satellite Service (FSS) transmissions currently occupy 5.9 to 6.4 GHz and 14 to 14.5 GHz in their uplinks, and 3.7 to 4.2 GHz and 11.7 to 12.2 GHz in their downlinks. Although the Direct Broadcast System (DBS) has been assigned 17.3-17.8 GHz for the uplink and 12.2 to 12.7 GHz for the downlink, not one of these DBS-U.S. high-powered satellites has been launched. As of now, there are 61 orbit locations assigned the U.S. in C and Ku bands.

With that history as background, let's first look at a television channel and then a series of satellite transponders that we'll determine both carrier-to-noise and signal-to-noise ratios, coupled with a general discussion of both signals. Following this, digital signals will be attempted, and then some audio measurements. All of these could add up to a fairly comprehensive chapter on servicing, some of which may be useful later on as HDTV terrestrial and satellite transmissions spread. Whether one or two TV channels, however, HDTV terrestrial will occupy no more than 6 MHz/channel, and satellite transmissions can have only a maximum of 500 MHz bandwidth/satellite, according to FCC regulations. Individual transponders at Ku, however, can occupy passbands to 72 and even 120 MHz, while C-band is restricted to 36 MHz plus guardbands.

Terrestrial TV is always first in our evaluations since HDTV authorization by the FCC depends largely on how the broadcasters put it on the air. Working in our accustomed power measurement of dBm (dB referenced to 1 mV), channel 4 video and audio carriers are 67.25 MHz and 71.75 MHz, respectively. In Fig. 8-18 readouts at a resolution bandwidth of 500 kHz, a center frequency of 68.88 MHz, and 0 dB attenuation, the 2710 marker shines brightly at 67.28 MHz for video, with audio just 4.5 MHz separated to the right. The next carrier visible to the right is that of channel 5. As you can see, the video marker's level is −39.3 dBm. Were we to rotate the antenna somewhat, this signal strength would increase, as will the carrier-to-noise (C/N) which we'll measure next.

Releasing the video filter, which adds another 4 dB to the carrier's amplitude, and energizing the Applications Menu for normalized C/N at 4.2 MHz, the analyzer automatically reads out 29.2 dB, with all other parameters the same (Fig. 8-19). If you want the signal-to-noise (S/N) measurement, simply execute this simple equation and add the result to C/N:

$$S/N = C/N + 20 \log 6/4$$

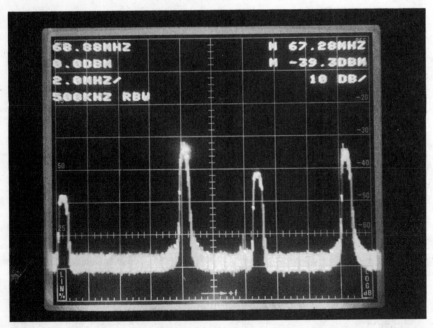

Fig. 8-18. Tektronix 2710 automated readouts on TV channel 4 using video filter—a delightful spectrum analyzer.

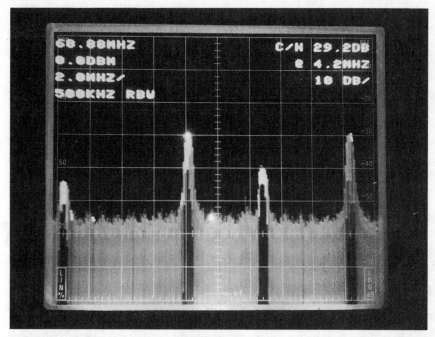

Fig. 8-19. Carrier-to-noise (C/N) auto readout with only a 2-button prompt. C/N reads out 29.2 dB top right at a center frequency of 68.88 MHz.

(6/4 being channel and video bandwidths. Therefore,

$$S/N = 29.2 \text{ dB} + 3.52 = 32.72 \text{ dB}$$

A C/N of 40 dB or better would have been considerably more desirable, but that's our analyzer readout following automatic instrument calibration, so we'll have to accept it as an accurate signal. Regardless, this is a good indication of what a modern spectrum analyzer can do when fully microprocessor controlled.

Satellite TV, however, is a dog of a different breed. In Fig. 8-20 you're looking at a total of four transponders on GE (used to be RCA) K1. The three center carriers are leveled at approximately −40 dBm, which is our requirement for a decent picture. The transponder on the left is down several dB more, perhaps due to fine tuning, a possible polarity shift, reduced power, or source origination. The one we're concerned with, though, is transponder 12 at 1243 MHz, almost exactly in the center of the graticule, whose frequency reads out at 1237 MHz.

This one is exactly −39.8 dB down according to the bright marker supplied by the 2710. On the baseband input of Toshiba's CZ2094 digital receiver, this level supplies a sharp, well-colored picture, with superior definition and resolution. We won't make comparisons with either terrestrial or cable TV emissions, but this is a very pleasing picture. The video filter is obviously not engaged.

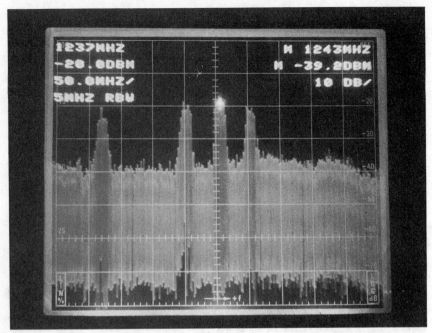

Fig. 8-20. Same analyzer, but this time reading satellite transponders following the block downconverter. These are active on G.E.'s K-1 spacecraft at about a −40 dBm level.

Here again we would like to know the carrier-to-noise readout of this signal taken directly off the Ku block downconverter. And this time the video filter *is* engaged (Fig. 8-21). We made an interesting discovery, in older spectrum analyzers, a correction factor had to be applied in FM satellite reception that amounted to roughly 12 dB. Here we're viewing the same 4 MHz signal as in analog video, with a resolution bandpass of 500 kHz (for this analyzer), so that the C/N must take this into account as well as the addition of 2.5 dB for the older analyzer log amplifier and video detectors. Consequently, you need an equation for the oldsters of:

$$C/N = 10 \log (4/0.5) + 2.5 = 11.53 \text{ dB}$$

which we'll round off to 12 dB. Now in our 2710, all you have to do is re-insert the video filter and push buttons for C/N, which reads out automatically to

$$C/N \text{ satellite} = 24.0 \text{ dB}$$

which you regard as almost identically equivalent to the full display with NO video filter inserted and 12 dB subtracted from C/N under those conditions. Isn't this procedure considerably easier than estimating the center of the noise level and then applying a general C/N correction?

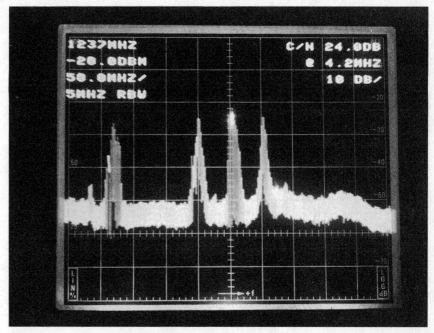

Fig. 8-21. A reasonably accurate, quick readout of C/N, with video filter inserted. The 2710 does the rest, then add 37.5 dB for S/N.

Say you want signal-to-noise (S/N), too, with all the rest? That's easy, too.

Satellite video S/N = C/N + 37.5 dB, or 61.5 dB

Is it any wonder you have a super clean picture? Studio quality is only 56 dB in the satellite business. So with good antennas, better than average block down-converters, a horizon-to-horizon mount that accesses satellites within 0.13 degree, U.L.-approved cable, and a good receiver with at least 27 MHz i-f bandpass, you're in business. Without these goodies you're in *deep* Ku band trouble, and as occupied bands ascend even higher, antenna tolerances, block downconverter noise figures, and antenna-to-receiver cabling specifications will tighten even more. Lacking a spectrum analyzer, transponder identification and measurements will be unadulterated guessing—and customers won't stand for that!

BASEBAND

Before AM or FM signals are modulated, their range of frequencies is called baseband. That term in the past few years, however, has become more collectively used until some, including your author, refer to general video and audio analog information as baseband. Some may disagree with this broad usage, but times are changing and there are lots of AM/FM signals meandering about that have to be called something. So baseband it is!

We will, however, introduce a new aspect to these modulating frequencies so you can see them better, and appreciate more up-to-date technology: it's called a digital storage oscilloscope (DSO) in a new version by none other than our old friend Tektronix.

DSOs

Relatively low cost, the Model 2210 (Fig. 8-22) sells for under $2500, has a 50 MHz bandpass, vertical deflection of 5 millivolts to 5 volts per division, with magnification to 500 μV/div. A variable volts/gain control can increase vertical deflection to 2.5 × vertical attenuator settings. The usual analog dual trace controls permit routine analog signal viewing, while digital bandwidth amounts to 10 MHz, rather than 50 MHz. Maximum sampling rates are 20 megasamples/second, and the waveform record accommodates 4 kilobytes for single and dual channel reproductions at a resolution of 400 points/div. Once stored in 8K memory, signals can be recalled for new signal comparisons as stored or magnified 10 to 50 times. Step response is 7 nsec or less, with maximum accuracy specified between +15 to +35°C.

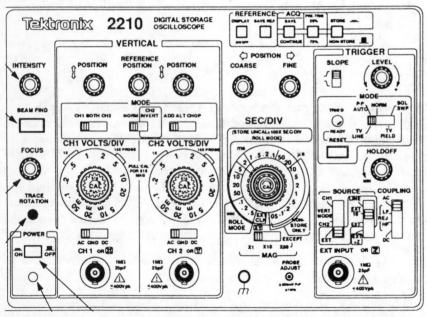

Fig. 8-22. Front panel drawing of the 2210, Tek's low cost 50 MHz bandwidth analog and 20 megasamples digital (DSO) low-cost oscilloscope with 8-bit vertical resolution and a 3-year warranty (courtesy Tektronix, Inc.).

Dual "state machines" in the digital section supply information to either the RECORD or ROLL modes. In RECORD, data is recorded and its starting address point loaded into the Address Counter and then transferred to the Display RAM (random access memory). In ROLL, information is displayed directly from the Acquisition RAM, clocked from the Display Controller. Updating in RECORD follows data acquisition, and while in ROLL, trigger signals don't count unless in Single Sweep. Otherwise, the CRT display is updated continuously at the digitizing rate. When Single Sweep operates, the last RECORD(ed) incoming waveform is displayed; and in ROLL, the display is continuous until the trigger circuit is armed and the pretrigger part of RECORD is full. A trigger-ready lamp then lights when data is acquired as RESET arms the trigger circuit. All this takes place in ROLL in realtime, very much like that collected in a strip chart recorder. For this particular equipment, keep digital input rates at 8 MHz or less, sampling drops to less than $2\times$ the period.

There are many other things to know about digital storage oscilloscopes (DSOs), especially the higher the investigated frequency the more samples and sampling rates are required. But with the complexity of consumer products increasing monthly, especially with digital logic and components, blips and glitches will cause many undesirable problems as this equipment ages. Therefore, DSOs are most important in capturing and displaying transients, single-shot events, pre-triggering, general 1s and 0s processing, and storage/displays for immediate or time-later analysis. Can your analog scope see and retain a fast glitch that's intermittent? Let's be realistic! (See Fig. 8-23.)

With just a little indoctrination and perseverance, this 2210 DSO isn't difficult to handle. In its basic state, we simply found a test point with the analog portion, pushed the storage button and, with only a slight level adjustment, the signal is there to be viewed at leisure or photographed. No serious attempt was made to explore more intricate measurements other than those immediately available at several test stakes on Zenith's module boards. It would appear that a DSO is the initial answer to testing digital equipment of the consumer variety, and we can afford to wait a little longer for a subsystem or system signature analyzer. The latter will come, but from initial explorations it isn't needed just now.

The photo in Fig. 8-24 is indicative of readouts in the digital portion of the receiver. In this instance, waveforms appeared at a 50 μsec/div. setting of the time base and at several volts amplitude. The top trace illustrates three rectangular pulse durations, while the bottom trace reveals a series of common serial excitations divided by an intervening "low" that blanks an interval of no information.

In Fig. 8-25, we just poked around the tuner's control section until we found some active, but rather ragged information in the mid-millivolt range at 20 μsec/div. As you can tell by the horizontal divisions occupied, some of this seems to be "klunking along" at a line sweep rate, but with prominent, but very rapid, transients peaking in between at amplitudes of some 0.8 V. In some instances, if these were pickups from line sweep they could disrupt logic counts

BASIC BLOCK DIAGRAM
DIGITAL STORAGE

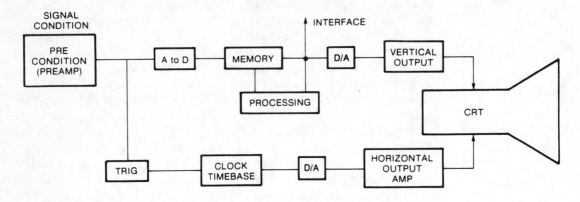

CRT STORAGE

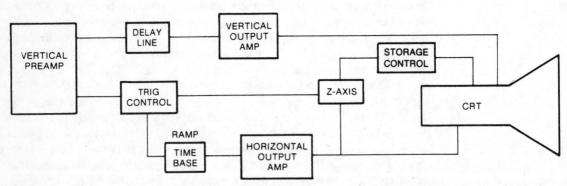

Fig. 8-23. Basic block diagram of 2210 illustrates two types of storage: CRT storage requires sophisticated cathode ray tube plus control circuits; digital storage requires A/D converters as well as memory, time base, and processing circuits (courtesy Tektronix, Inc.).

all over the place. Without a schematic we really don't know, but what is becoming more and more evident, the probability of using DSOs in both signal processing and microprocessor and tuner command signals is very apparent. Storage features, plus one-shot capabilities seem to do the trick every time. Sorry we can't devote more time to this digital "adventure," but the chapter involves only basic applications and is not an in-depth study.

MTS AUDIO

As promised, the final portion of the chapter will talk briefly about audio and illustrate the explanation with some low frequency spectrum analyzer readouts.

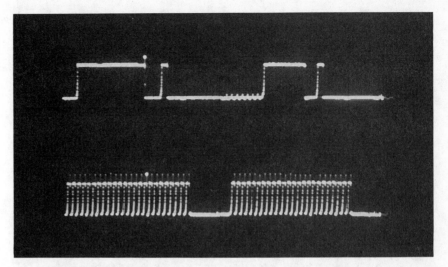

Fig. 8-24. Dual exposures of stored signals from Zenith's digital converter.

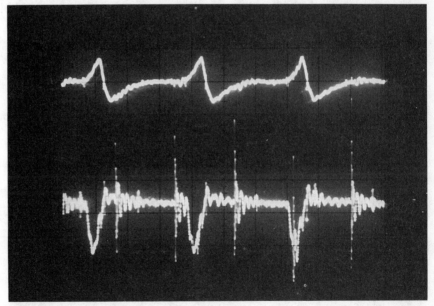

Fig. 8-25. Poking around in the tuner control produced these responses that can only be displayed by using both storage and single sweep.

Multichannel sound, or BTSC-dbx, was finally approved by the Federal Communications Commission on March 29, 1984 and is the *only* stereo sound system permitted on broadcast television. The main channel identifies as L + R, while a subcarrier left channel appears as L − R, is level encoded and produced as double sideband, suppressed carrier AM modulation at 2_H, or double the rate of the pilot frequency, identical to our NTSC scanning rate of 15,734 Hz. Due to cost and somewhat unexciting results, MTS is by no means universal throughout the broadcasting industry, with some 45% of all TV stations so equipped by 1989. The spread of HDTV in its early stages may even be somewhat slower because of combined transmitter modifications and initial receiver costs. But the digital sound that's promised with HDTV could and should be most welcome if detected and processed with suitable decoders, amplifiers, and speakers—all of which will vastly exceed the poor-quality $299 television sets of today.

Using an industry-approved MTS 2009 B & K-Precision stereo generator and a color bar generator with external audio input, let's briefly examine the stereo product of Zenith's digital receiver and see just how good it is. With a considerably improved sound system that's already mainly digital, the results should be interesting (Fig. 8-26).

Good, but not as spectacular as promised. The 7L5 (dc to 5 MHz) analyzer shows response between 400 Hz and 13 kHz to be essentially flat. After that, however, the dropoff is at least 20 dB at 14.5 kHz, followed by a horizontal flyback blip at 15 kHz. After this you see a rise in frequency toward 20 kHz that using a higher frequency signal input we could not duplicate. Therefore, we'll say that cutoff comes at 14.5 kHz, which should be enough for anyone. The top readout, you may notice, is – 8dBV; the reference being 1 volt.

The next thing you might like to know about TV stereo is channel separation. This is done with a right or left signal from the MTS 2009 and the differ-

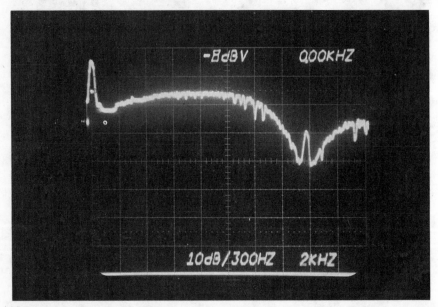

Fig. 8-26. Essentially flat to 14.5 kHz, Zenith's digital TV's audio frequency drops off rapidly after 14.5 kHz. Even superior ears won't notice.

ence between L and R receiver outputs. In this instance (Fig. 8-27) we had 1.4 volts in the lower trace versus 12 volts in the upper trace. So we find:

20 log 12/1.4 = 20 log 8.56 = 18.66 dB = R/L channel separation

Finally, such measurements are not complete unless you check harmonic distortion. Once you know how and have a handy table (available from Tektronix and probably others), HD is easily calculated by adding all harmonics together that aren't 10 dB different from the 2nd. Now the separation between fundamental and 2nd harmonic measures some 37 dB, while the 3rd harmonic is yet another 3 dB down; and the 4th, being 10 dB from the 2nd doesn't count. So, according to the table, a 3 dB harmonic difference between 2nd and 3rd causes 1.76 to be added to 37 dB, making to total 38.76 dB, or a distortion of 1.15%. Not too bad, considering the combination of analog and digital processing (Fig. 8-28).

DIGITAL MULTIMETERS

Both new in function and concept, the FLUKE 80 SERIES appears as a brand new and very worthy successor to the 70 series of years past. Retaining auto calibration and important analog and digital functions, this new series, especially Model 87 (Fig. 8-29), can now measure frequencies between 0.5 Hz and 200 kHz with up to 0.01 Hz resolution. Changes as short as 1 msec can be

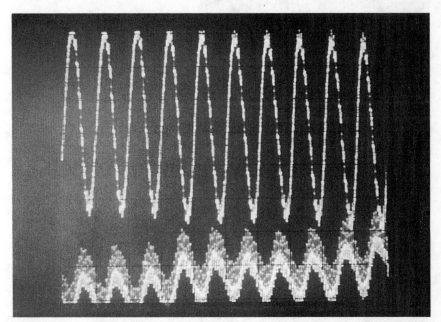

Fig. 8-27. Channel separation of 18.66 dB is easily computed with DSO storage control and dual traces.

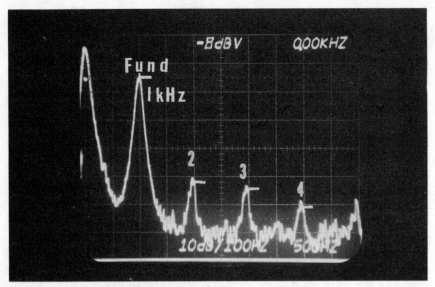

Fig. 8-28. This waveform is exhibited by an older 7L5 analyzer showing harmonic distortion of 1.15% when the 2nd and 3rd harmonics are specially summed and subtracted from the fundamental at 1 kHz.

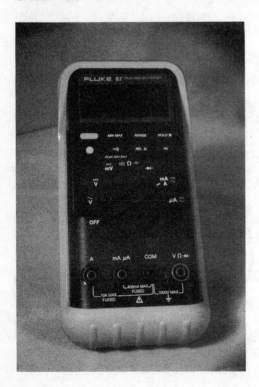

Fig. 8-29. Fluke's new 87, microcomputer driven 4.5 digit multimeter that has all the usual, plus capacitance and frequency measurements to 200 kHz, waveform duty cycles, auto min/max and hold (courtesy John Fluke Mfg. Co.).

recorded, and an alternate frequency counter can recognize the length of a duty cycle and read values between 0 and 99.9%. Readouts can also be held, and a separate mode permits initial measurement storage and then displays the difference between this and succeeding readouts. A 4.5 digit display also increases resolution $10\times$ over 3.5 digit units. And if you plug test leads into the wrong input terminals, the meter alarm beeps to warn you.

If that isn't already enough, there's a blue button that will allow toggling between ac or dc in μA, mA, or ampere functions, or ohms and capacitance, and capacitors can be evaluated from 5 nanofarads (nF) to 4 microfarads (μF). Continuity examinations are also available with a threshold of 10% of each range and a meter audible beep, and should you really want to record minimum and maximum readings up to 36 hours can be stored in memory, and the results toggled for instantaneous evaluation. While you will want to use auto ranging most of the time, there is a Range switch for manual settings, while a return to auto takes a Range press of 2 seconds. The same time also exits Min/Max and erases prior readings.

Theory of Operation

Fluke's own block diagram just issued, appears in Fig. 8-30. Note that the instrument is divided into analog and digital divisions, with IC chip U4 handling both operations via its microcomputer. Here, U4 executes software commands, does data formatting, and selects various digital and analog operations, accompanied by LCD readouts and audible noise beeps. The digital U4 segment furnishes synchronous A/D control, with a 16-bit counter monitoring counts and frequency measurements. There are bus and interrupt control circuits, as well as registers for analog switch drives. The volts/ohms and diode test input is overload protected by a pair of MOV (metal oxide varistors), three current-limiting resistors, and a spark gap, while the mA and μA inputs and the 10 ampere input are fuse and resistor-protected.

Inputs pass through these protect circuits and on to a double-sided wafer switch for function selections. At turn on, IC U4 exercises a voltage ratio that identifies it or them. Compensation and scaling then takes place, followed by A/D conversion, autoranging, comparison, etc., in the microcomputer. A/D conversion includes both dual rate and dual slope with a basic rate of 40 measurements/second. A full 4k count delivers full-scale measurement, along with sampled data for bar graph analog display. New samples are accumulated every 25 milliseconds, and 8 are summed for full-resolution readout. Basic A/D timing is set for eight read cycles and then a 40 msec autozero phase, with some variations for special operations. Capacitors are measured by the *charge* needed to change some potential across the unknown from 0 to system reference voltage, with two samples for each display update.

In the digital portion, the A/D converter is managed by a 4-bit state machine with ROM output decoding; a 16-bit counter "reads" A/D conversions; a 48-bit write-only register latches microcomputer information for analog

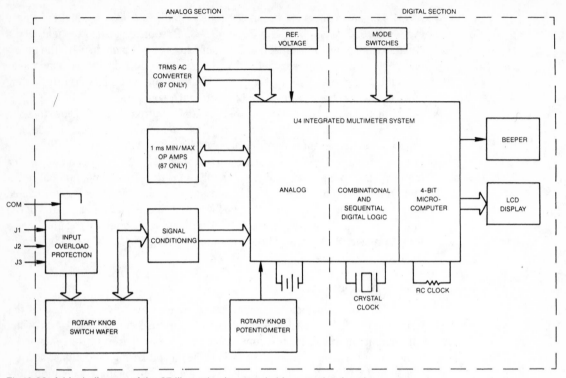

Fig. 8-30. A block diagram of the 87 illustrating its remarkable operating functions (courtesy John Fluke Mfg. Co.).

switch drive; an interrupt control cell multiplexes and manages four interrupts to the microcomputer that are for the 16- and 8-bit counter carries and voltage comparator transitions; a bus control does the address decoding and power down, jack sense, and weak battery warnings.

In other operations, the 16-bit counter is divided into two 8-bit counters, counts are accumulated in extended 20-bit registers and the unknown frequency is calculated from counter value ratios. Start/stop measurement cycles are controlled by the microcomputer, and the value ratios describe the duty ratio.

The 4-bit microcomputer housekeeps instrument operations and displays. It has its own clock, which cycles between 525 and 800 kHz. In frequency measurements, the microcomputer starts the counters, checks count accumulations, arms and disarms signal gating logic, and calculates the final display. Analog switches are directly controlled, and in auto-ranging, proper ranges are determined from signal inputs. Power supplies consist of one shunt and one series regulator.

Min/Max operations result from operational amplifier, diode, and capacitor charge actions for positive and negative signal excursions, which are stored and then read out on the display with minimum and maximum values.

Instrument ranges on ac are from 0.1 mV to 1,000 V; dc from 0.01 mV to 1,000 V; ohms, 0.01 to 40 megohms; capacitance, 0.01 nF to 5 μF; current, 0.1 μA to 400 mA/600V, and dc current to 10 amperes.

CONCLUSIONS

The preceding should have at least offered a few ideas on what's needed in the digital/HDTV service routines we will all be investigating in the coming years. New technologies require new approaches, different and often more exotic instrumentation, additional technical training, and an open mind toward getting the job done. The individuals most willing to adapt and move ahead will remain winners in both business and at the cash register.

Like the fast-disappearing Mom and Pop stores who have depended on a little service and some sales to keep them going, even the medium-sized retailers will soon have to face a service crisis of considerable proportions or enter another line of endeavor. All sorts of products bring wonderful entertainment and freedom from drudgery, but they eventually do break down, and someone has to be around to refurbish and rebuild.

About the best way we know to intelligently react is to read and investigate all the various possibilities, preferably *before they arise*, rather than after they're here. A serious case in point are the thousands of satellite earth terminals around the country, having been sold several years past they have virtually no one to service them. Some of you readers may seriously want to enter the business if satellite owners can afford to pay the bills. And should system replacement become necessary, do give the commercial or consumer owner a real break with first-class equipment and installation – poor people, they deserve it!

The Boy Scout motto of "be prepared" never applied more realistically to the service industry than it does today. And "them that has, gets," while those without are lost. Do be prepared as HDTV and digital logic in all forms rises above the horizon in this spring of advanced communications. We'll try and help!

9

The Chairman's Report to the 101st Congress

THE FOLLOWING REPORT BY FCC ADVISORY COMMITTEE CHAIRMAN RICHARD E. Wiley on behalf of the Advanced Television Service to the U.S. House of Representatives Subcommittee on Telecommunications and Finance is considered an extremely important document in the search for a workable U.S. high definition TV system (HDTV) and, because of its contents and complete public disclosure, is presented in full so that additional insight into this multifaceted problem seeking a solution is published for all to see. As reported earlier, attorney Wiley has also been Chairman of the Federal Communications Commission from 1974 to 1977 after service in that agency as both General Counsel and Commissioner, beginning in 1970.

The Advisory Committee on ATS was established by the Commission in 1987 for policy, standards, and regulation advice, and this report is an overview of consumer electronics in the U.S. economy, assessing the potential impact on our domestic consumer electronics industry, and offers certain policy guidelines. Committee members, according to Mr. Wiley, include TV broadcast networks, television stations, equipment manufacturers, CATV representations and their "diverse" views. Later reports are expected from the Committee during 1989 and 1990 as testing and hardware evaluations are received from other ATS members.

HDTV ISSUES

The potential development of Advanced Television Service (ATV) or High Definition Television Service (HDTV) has sparked interest in a revival of the U.S. consumer electronics industry, and in the potential impact of such a revival on this country's technological base. Because a large share of the current U.S. consumer electronics industry is foreign owned, there are strong differences of opinion among U.S. market participants concerning the appropriate U.S. government policies for developing ATV so as to provide the maximum benefits to the U.S. economy—both to consumers and producers. Most of the current debate involves the importance of U.S. research and development in ATV and the emergence of a U.S. ATV standard that embraces a technology owned by U.S. firms.

Some industry participants believe that the ATV technology used in the United States must be developed and licensed by U.S. firms if the full benefits of ATV production are to be shared among U.S. manufacturers of semiconductors, high-definition receivers, and other components. They believe that the research development involved in bringing this technology to market will have important effects on other products, and industries in this country, and even on U.S. national security.

Others argue that U.S. participation should not be measured on the basis of U.S. ownership of technology or facilities, but rather in terms of the contribution of that technology to U.S. consumers and owners of productive resources, including labor and capital. The latter argue that the U.S. economy benefits from domestic production whether the new technology is controlled by foreign-owned companies or U.S.-owned firms as long as a large share of the value added in research, development, and manufacturing takes place in the United States.

It is impossible in the time allotted for this report, and with the resources available to the Advisory Committee, to provide more than an overview of the difficult issues that have arisen. The number of issues that eventually must be analyzed is very large, but the most important may be identified as follows:

- Should the U.S. government stimulate domestic research and development in ATV through direct support?
- Should a consortium of potential U.S. producers be formed, with the support of government (and perhaps even a waiver of antitrust restrictions), to develop ATV?
- Is a U.S.-owned technology, with patents held by U.S. firms, essential to the growth of the semiconductor or electronics industries in the United States?
- Is full HDTV quality essential for the production of ATV receivers in the United States?
- Will ATV receiver manufacturers in the United States be influenced by U.S. or foreign ownership of the underlying technology?
- Is the revival of a U.S.-owned consumer electronics industry possible, and is ATV the vehicle to achieve such a revival?
- Will ATV technology be crucial to the future development of U.S. manufacture of other high-technology industries such as personal computers or automated manufacturing equipment?

Obviously, these questions cannot be answered in a few months through the deliberations of industry participants. Rather, appropriate policy judgments could better be developed through detailed market surveys and economic analyses of the various issues. In this report, we provide the best evidence available to the Committee on a number of these issues, but we do not suggest that this evidence is sufficient to yield firm policy conclusions.

THE CURRENT MARKET

Television Receivers. Factory sales of television receivers in the United States are projected to be $7.2 billion in 1989. This will represent about 20 percent of the factory value of all consumer electronics sold in the United States (Table 9-1). Roughly two-thirds of all sets sold in the United States are assembled in domestic plants. These U.S.-produced sets account for nearly 80 percent of factory sales of television receivers because imports are most highly concentrated in the smaller sets.

A recent draft report of the Electronic Industries Association (EIA) concludes that the total domestic content in U.S.-assembled sets is about 72 percent of factory cost or about 85 percent of retail value. According to this same EIA study, virtually all sets with picture tubes bigger than 20 inches are assembled in the United States. Therefore, as U.S. buyers migrate to larger and larger sets, the average domestic content in NTSC receivers sold in the U.S. slowly rises.

In 1989, total U.S. retail sales of television receivers will be $10.2 billion, of which $8.0 billion will reflect sales of U.S.-produced receivers, according to EIA. The EIA analysis predicts that U.S. production and distribution will contribute $7.4 billion of the $10.2 billion total economic value of these sets (Table 9-2).

Of course, a very large share of U.S. receiver production is now conducted in foreign-owned plants. There is only one major U.S. company left in television receiver manufacturing, Zenith, with only 12.8 percent of the U.S. market, and much of its manufacturing now takes place abroad. RCA, now owned by Thomson, has 22.2 percent of the domestic market (including both GE and RCA sets), and a number of Japanese, European, and Korean firms account for most of the rest. Thomson/RCA, Philips, Toshiba, Matsushita, Sanyo, Mitsubishi, Sony, Samsung, Goldstar and Sharp now account for a large share of U.S. production of television receivers (Table 9-3).

The Role of Semiconductors. According to the American Electronics Association (AEA), NTSC television receivers sold in the United States will account for $664 million in semiconductor sales in 1990. This represents about $29 per current receiver, or about 7.5 percent of the value of the receiver, far less than the potential value of semiconductors that will be required in prospective ATV receivers. The AEA report does not identify the U.S. semiconductor manufacturers' participation in the current market for NTSC receivers. However, even in the unlikely event that all sets currently assembled in the United

Table 9-1. Total Factory Sales of Consumer Electronics by Product (Millions of Dollars).

Product	1985	1987	1989
VIDEO:			
Color TV (including LCD)	5,562	6,271	6,405
Monochrome TV (including LCD)	309	287	249
Projection TV	488	527	515
Total VCRs	4,738	5,093	5,085
Video Disc Players	23	30	63
Video Cassette Players	22	26	29
Home Satellite Systems	900	625	675
Total Video Products	12,042	12,859	13,021
AUDIO:			
Audio Systems	1,372	1,048	1,065
Separate Audio Components	1,132	1,400	1,500
Portable Audio Tape Equipment	1,140	1,431	1,350
Home Radios	379	409	390
Total Autosound Equipment	3,000	3,800	4,400
Total Audio Products	7,023	8,088	8,705
HOME INFORMATION EQUIPMENT:			
Home Computers	2,050	2,920	3,500
Corded and Cordless Telephones	970	1,055	1,095
Cellular Telephones	115	265	310
Telephone Answering Devices	266	410	490
RELATED PRODUCTS:			
Cassettes, Accessories	3,333	4,670	6,100
Home Security Systems	600	800	1,000
GRAND TOTAL	26,399	31,067	34,221

Source: Electronic Industries Association.

States contained only U.S.-produced semiconductors, total U.S. semiconductor sales to this market would be less than $600 million in 1989.

Other Components. The EIA analysis of the contributions of the current color television industry to the U.S. economy suggests that the domestic content of all U.S.-assembled color television receivers is about 79 percent of factory value. Given that value-added in the television receiver industry itself is between 30 and 40 percent of factory value, the remaining value added, such as that provided by components and cabinetry, must represent 39 to 49 percent of total value or between $2.2 billion and $2.7 billion per year.

In a preliminary analysis of the potential for HDTV manufacture, EIA estimates that television manufacturing accounts for 45,000 jobs in the United States and other manufacturing jobs required for the production of receivers (including components) account for 41,000 jobs. Thus, nearly half of the domestic employment accounted for by television set manufacture appears to be in upstream components industries.

Table 9-2. Domestic Content of Television Receivers*, 1989 Estimates (Billions of Dollars).

	DOMESTIC PRODUCTION	IMPORTS	TOTAL
Factory Value	5.6	1.6	7.2
U.S. Value Added in Television Industry	1.7-2.2	0	2.2
U.S. Value Added in Components, Cabinets, etc.	2.2-2.7	0	2.2
U.S. Value Added in Distribution	2.4	0.6	3.0
Total Domestic Content	6.8	0.6	7.4
Retail Value	8.0	2.2	10.2
Percent Domestic Content	85%	27%	73%

*Note: Includes all monochrome, color, and projection TV sets.

Source: Electronic Industries Association.

Table 9-3. U.S. Set and Tube Production (1988 Data).

Company	Location	No. of U.S. Employees	Plant Type	Annual Production
BANG & OLUFSEN (Joint venture with Hitachi)	Compton, CA (opening 1989)	(See Hitachi below)	Assembly	Not Available
GOLDSTAR	Huntsville, AL	400	Total Production	1 million
HARVEY INDUSTRIES	Athens, TX	900	Cabinet Assembly/ TV Assembly	600,000 capacity
HITACHI	Anaheim, CA	900	Total Production for 24", 27", 31"	Over 360,000

Table 9-3 continued.

Company	Location	No. of U.S. Employees	Plant Type	Annual Production
JVC	Elmwood Park, NJ	100	Total	480,000
MATSUSHITA	Franklin Park, IL	800	Assembly	1 million capacity
AMERICAN KOTOBUKI	Vancouver, WA	200	VCR/TV Assembly	Not Available
MATSUSHITA (Joint venture w/Philips)	Troy, OH (opening Spring 1989)	1-200 upon opening; eventually 400	Tubes	1 million
MITSUBISHI	Santa Ana, CA	550	Final Assembly	400,000
MITSUBISHI	Braselton, GA	300	Total/Full	285,000
NEC	McDonnough, GA	400	Final Assembly	240,000
O-I/NEG TV PRODUCTS (Joint venture of Owens-Illinois & Nippon Electric Glass)	Columbus, OH	800	Tubes (Capabilities) up to 45″)	Not Available
O-I NEG TV PRODUCTS	Pittston, PA	750	Tubes (Capabilities up to 45″)	Not Available
O-I/NEG TV PRODUCTS	Perrysburg, OH	75	Component glass for TV (soder glass)	Not Available
ORION	Princeton, IN	250	Assembly	Not Available
PHILIPS	Arden, NC	4-500	Parts (Plastic Cabinets & Accessories	Not Applicable
PHILIPS	Greenville, TN	3,200	Assembly /Full	Over 2 million
PHILIPS	Jefferson City, TN	1,000	Parts (Wood Cabinets)	6 – 700,000
PHILIPS	Ottawa, OH & Emporium, PA	2,300	Tubes	3 million
SAMSUNG	Saddlebrook, NJ	250	Production for 13″- 26″ TVs	1 million capacity
SANYO	Forrest City, AR	400	Assembly	1 million capacity
SHARP	Memphis, TN	770	Assembly	1.1 million
SONY	San Diego, CA	1,500	Full Manufacturing of Color TVs & Tubes	1 million

Table 9-3. Continued.

Company	Location	No. of U.S. Employees	Plant Type	Annual Production
TATUNG	Long Beach, CA	130	Assembly	17,500
THOMSON	Bloomington, IN	1,766	Full Manufacturing/ Assembly	Over 3 million
THOMSON	Indianapolis, IN	1,604	Components Manufacturing (Printed Boards & Cabinet Production)	Not Applicable
THOMSON	Mocksville, NC	626	Cabinet Production	Not Applicable
THOMSON	Marion, IN	1,982	Tubes	Not Available
THOMSON	Circleville, OH	700	Glass for Picture Tubes	Not Available
THOMSON	Scranton, PA	1,242	Tubes	Not Available
TOSHIBA	Lebanon, TN	600 (300 add'l planned for 1989)	Assembly	900,000 (1.3 million planned for 1989)
TOSHIBA	Horseheads, NY	1,000 (500 add'l planned for 1989)	Tubes	1 million (1.5–2 million planned for 1989)
ZENITH	Springfield, MO	2,000–2,500	Full Manufacturing/ Final Assembly	Not Available
ZENITH	Melrose Park, IL	2,500–3,000	Tubes	Not Available
TOTALS: 20 companies	35 cities or plant locations in 15 states	32,695 production employees	Not Applicable	Annual production of television receivers: 16,482,500+ (figure does not include tube production)

THE PROSPECTIVE IMPACT OF ATV

The debate over U.S. policy involves far more than its impact upon the future of television receiver manufacture in the United States. Rather, it is a debate about how to stimulate research and development on important new technologies in this country. It is a debate about the importance of ATV or HDTV research in particular and electronics research in general on the U.S. industrial base. And it is a debate about the potential role of breakthroughs in ATV or HDTV technology on other industries, including some with national-defense implications.

Research and Development. There is clearly a concern that research in consumer electronics has shifted away from the United States. At present, only three consumer electronics firms (Philips, Zenith and Sarnoff) have major research facilities in the United States, and two of these are now foreign-owned. ATV could provide the opportunity to expand these activities and even to admit new entrants into U.S. consumer electronics research. Obviously, a strong research capability for consumer applications of ATV could well yield benefits in other products. Moreover, this capability would add to the United States' general technological capacity.

There are some who also argue that the flow of patent monies could be important to U.S. electronics firms, but these patent revenues are likely to be small relative to the other benefits of ATV development. Moreover, these patent yields may be small relative to the capital invested in research—a fact consistent with the observed reluctance of many U.S. firms to invest in this activity.

Television Receivers. Obviously, the importance of ATV to the U.S. economy will depend very much on the rate at which sets are purchased by U.S. consumers regardless of the technology finally employed. A 1988 report prepared for The National Telecommunications and Information Administration predicts that 4.1 million ATV sets will be produced in the year 2000. More recently, the EIA has estimated that 10.4 million HDTV sets will be sold annually by the year 2000. The AEA also assumes that 10 million sets will be sold per year by 2000, but 90 percent of these will be some form of "enhanced definition television" (EDTV) rather than full HDTV. For this reason, AEA forecasts a U.S. retail market value of ATV receivers of $9.6 billion in 2000 while the EIA forecast, based upon a somewhat greater penetration of higher-quality ATV receivers, forecasts an $18.5 billion market.

The EIA report predicts that 10.2 million of the 10.4 million HDTV sets sold in the United States in 2000 will be assembled in the U.S. and that domestic content will account for 85 percent of the *retail* value of these sets. This conclusion is based only upon the observation that large color NTSC sets are now largely U.S.-produced. Neither EIA nor AEA provides a detailed analysis of the determinants of manufacturing for HDTV sets, presumably because the technology of these sets is still largely unknown. Obviously, all estimates of ATV receiver production and sales must be viewed as rough estimates at this juncture, given the current technological and regulatory uncertainties.

Other Electronic Components. The EIA forecasts that $15.7 billion of the $18.5 billion market for HDTV in 2000 will reflect U.S. value added. The remainder consists of imported electronic components. AEA, more concerned about the effects of HDTV on semiconductors, stresses the fact that HDTV production for the U.S. market could contribute nearly $700 million per year to semiconductor sales. If, however, the EIA forecast of HDTV receiver sales is correct, semiconductors sales to this market could exceed $1 billion annually.

Other Industries. The AEA assumes that a strong U.S. HDTV and semi-conductor industry will redound to the advantage of U.S. production of personal computers and automated manufacturing equipment. This assumption is based on the theory that U.S. research and development in the areas of flat panel displays, high-speed memory devices, intelligent signal processing, and other related technologies will have large spillover effects on other industries.

AEA provides estimates that a strong U.S. HDTV industry will stimulate $42.5 billion of U.S. manufacturer sales of PCs and $3.5 billion of automated manufacturing equipment in the year 2000 over the level that would be achieved if the United States is "weak" in HDTV manufacture. Obviously, these are informed judgments of AEA members; at this juncture it is impossible to forecast the technological spillovers from U.S. HDTV development. Nor is it clear why U.S. manufacture of PCs and automated manufacturing equipment cannot thrive even if the U.S. ATV system is based on a foreign technology. Presumably, the reasoning is that a foreign ATV technology will stimulate demand for non-U.S. semiconductors and displays, and that U.S. semiconductor and electronic firms will thus fall behind these foreign producers.

An Assessment of the Current Television Receiver Market

The above analysis demonstrates that there is still a large television receiver industry in the United States, but that it is largely foreign owned (particularly since General Electric's sale of RCA's consumer electronics division to Thomson). Moreover, the receivers produced in the United States include foreign components and subassemblies that represent more than 20 percent of their factory value. It is therefore clear that the U.S. television receiver industry is part of a global industry that obtains technology, capital, and components from many foreign sources. As the dollar weakens and sets become larger, the domestic content in U.S.-assembled sets may stabilize and even increase slightly, but there is no denying that consumer electronics production and assembly is now an industry with a great deal of international specialization. Even if all U.S. television receivers were produced by U.S.-owned companies, the share of domestic content might remain at or below 79 percent of factory value because of this specialization.

At this juncture, no one has provided a definitive analysis of the effects of HDTV on U.S. consumer electronics manufacture. The EIA report is sanguine about U.S. participation in this market. The AEA report is plainly more worried. But neither can demonstrate dispositively that HDTV components, displays, or final receivers will not be produced in lower-wage developing countries. Obviously, given the uncertainties over HDTV technology, such an analysis would be exceedingly difficult. It is likely, however, that large HDTV receivers and displays will be produced in this country for U.S. consumption unless the dollar rises sharply to offset the transportation costs involved in importing these products.

POLICY OPTIONS

Presumably, the goal of U.S. communications policy should be to induce the development of an ATV service that makes a maximum contribution to U.S. consumers and producers and to the U.S. technology base (while, at the same time, preserving the public's interest in existing transmission technologies). If policies can be developed that allow for a substantial domestic contribution from semiconductor producers, other components suppliers, receiver manufacturers, or production and distribution equipment suppliers without unduly penalizing consumers, these policies should be considered. If, however, the search for policies to generate a revival of U.S.-owned consumer electronics manufacturers serves to unduly delay the development of an ATV service in the United States, such policies may be unwise and, indeed, futile. Undue delay in establishing the appropriate policy for terrestrial or satellite broadcasts of HDTV will provide the opportunity for foreign producers to establish their technology through alternative media—such as cable or VCRs.

There are a number of possible policies that have been suggested for ATV development. These include various degrees of government encouragement, support, or subsidy for research and development, manufacturing startups, or even consortia. A brief discussion of these follows.

Research and Development

There is considerable concern that the Unites States is falling behind other countries in various high technology sectors because of a decline in research and development outlays for civilian activities. U.S. government funding of R&D is largely concentrated on military projects while other countries, such as Japan, target civilian industries for such support.

Obviously, U.S. producers and consumers also benefit from foreign research. However, some U.S. interests express the belief that U.S. ownership of the ATV technology would confer a number of "first mover" advantages upon the U.S. electronics industry. First, U.S. ownership might provide a flow of patent license revenue—although the record is unclear as to how significant these royalties might be. Second, in light of the size of the U.S. market, it would create an opportunity for American companies to move along the progress curve somewhat ahead of their foreign rivals.

There is also the possibility that new HDTV technologies will have important national defense implications. For this reason, the Defense Advanced Research Projects Agency (DARPA) has recently announced its intention to fund major research into advanced television display technologies.

Before proceeding with any additional government support of ATV R&D, two important questions should be considered. First, is there reliable evidence that support of ATV research is likely to be more productive than support of numerous other emerging technologies? Second, is it likely that such support would pay dividends prior to the time that other countries, such as Japan, will have suppliers in production and be well down the learning curve in manufacturing ATV or HDTV equipment to their countries' standards? At the present time, the Advisory Committee does not have an adequate factual basis to independently draw reliable conclusions on either of these questions.

SUPPORT OF U.S. MANUFACTURING

Some industry participants believe that the U.S. environment is not conducive to large-scale risky investments with distant payoffs—such as those in the prospective ATV industry. They therefore suggest a variety of government policies for ameliorating this risk. Among the devices are consortia of prospective manufacturers which could jointly develop technology and manufacturing processes, thereby eliminating competitive uncertainties and reducing the individual company's capital requirements. One problem with the consortium approach is that it may require modification of the antitrust laws, and it might be difficult to argue that such a relaxation should be limited only to ATV. Thus, Congress may have to examine the wisdom of a general relaxation of the antitrust laws regarding horizontal arrangements such as consortia. Impetus for such an examination has recently been provided by the Secretary of Commerce in the Reagan Administration, William Verity, and Attorney General Thornburgh.

Another possibility is a government subsidy to capital invested in technology with long developmental lead times. Such industries as biotechnology, other pharmaceuticals, various capital goods, and even semiconductors might also benefit from such a subsidy. Obviously, Congress would have to determine which, of any of these industries, represent appropriate candidates for Federal funding.

DOMESTIC CONTENT

Some U.S. interests believe that the combination of an appropriate ATV standard and other policy initiatives may restore the domestic content to consumer electronics sold in this country. However, a substantial share of the cost of receivers assembled in the United States currently is embedded in imported subassemblies. The economics of this consumer electronics production may not be changed by ATV policies. Nevertheless, ATV receiver manufacture may involve large display units and a large quantity of semiconductors that can be produced efficiently in the United States. If, as anticipated, the U.S. represents a primary market for ATV, there is a good prospect that this production may develop in the United States.

However, it is at least questionable whether U.S. production of the gamut of consumer electronics would be restored simply by the prospective development of a U.S. ATV industry. Cameras, camcorders, VCRs, audio equipment, CB receivers, and various other consumer electronics products likely would continue to be manufactured or assembled in countries with lower wages or more efficient facilities than the United States.

CONCLUSION

There is no doubt that recent trends in U.S. productivity, civilian research and development, and savings rates have reduced U.S. industrial competitiveness. A weaker dollar has reduced the U.S. external trade deficit and has made U.S. manufacturing more competitive. However, Congress needs to address policies that will restore productivity growth and the general growth in U.S. living standards.

High definition television provides a major opportunity for U.S. electronics producers to develop technologies with potentially far-reaching benefits. The extent of these benefits and the pace at which they will unfold cannot be predicted at this time. However, Congress should be attentive to policy proposals that will increase the probability of success of U.S. firms in developing frontier technologies, such as those required for HDTV. Most importantly, the U.S. needs to expand its technology base through greater expenditures on research and development, increased private investment, and improved education. Policies designed to achieve each of these objectives should benefit U.S. electronics production in general and HDTV development in particular.

While it is currently unclear whether U.S. interests would be optimized by policies *specifically* designed to promote American involvement in ATV, the Advisory Committee believes that Congress should take a hard look at each of the policy options discussed above. In our opinion, ATV represents a potentially significant area of opportunity for the nation's technological base and the overall national economy. Accordingly, it would be a mistake to allow this opportunity to pass without giving careful consideration to the ways in which government and industry might cooperate to promote the maximum possible benefit to the American people.

This concludes Mr. Wiley's highly informative and reflective HDTV report to the Telecommunications and Finance Subcommittee of the U.S. House of Representatives. You may want to re-read some of its definitive descriptions. Almost everything we're working with these days is well laid-out and tentatively evaluated in preliminary analysis. To be sure, there are any number of problems facing the forthcoming HDTV industry, but not all are insurmountable by any means. In the final analysis, the greatest question probably revolves about our national will to do the job. We can IF we want to.

10
Preliminary Evaluations and Conclusions

AT THIS FORMATIVE MOMENT IN HDTV DEVELOPMENT TWO OLD ADAGES SEEM TO apply: "If wishes were horses, beggars would ride," and "Put your money where your mouth is."

As the end of the 20th century approaches and some governments are discovering how to manage booms and busts so neither is necessarily severe, the "wishers" in HDTV predominate, while money interests have a great deal to talk about but no overwhelming resolve for action. About the best indications available suggest that those European, Orientals, and one or several U.S. companies will continue to make any and all television receivers, and everyone else will gather 'round to assess the results. TV transmitters could be a somewhat different story, but standard or HDTV cameras in studios may largely remain Japanese. Video cassette or reel recorders could probably occupy the same category. So what else is new?

For now, at least, there is a great deal of industrial, government, and Congressional activity, with position papers flying everywhere, especially in Congress, and some of the commercial associations, particularly those who solicit Federal backing for development funds. What they'll receive is problematical, depending on technical promise, political connections, and the economy's purse strings. So far, great rhetoric, rally 'round the flag, make big plans, but the Congressional budget probably comes first. After that, perhaps there's a crumb or two remaining to motivate at least part of the process. If not, nature takes her course and the foreigners move in permanently.

Current Participants Are

The House Telecommunications and Finance Subcommittee; the Senate Commerce Committee; the Advanced Television Services Committee of the

Department of Commerce; the Electronic Industries Association; The National Association of Broadcasters; The Semiconductor Industry Association; the American Electronics Association; plus all the other Agencies and system proponents we've already discussed in previous chapters are current participants. Obviously, there is considerable interest, but only the ATSC group has offered earnest money for the Advanced Television Test Center. Are we really to expect more? Only time, public prodding, and the profit motive can tell.

In the meantime, we understand that 20 companies of the H-P and Tektronix types have put up $5,000 each to see what they can do. All belong to the American Electronics Association, headquartered in Washington, D.C. with a membership of some 3700 companies. Those 20 now involved include: Hewlett-Packard, Tektronix, Digital Equipment, ITT, Apple Computer, IBM, Motorola, Texas Instruments, Zenith, AT&T, AVX, Analog Devices, Cohu, Micro Electronics Computer Technology, Ovonic Imaging, PCO, Inc., Prometrics, Raychem Corp., The Grass Valley Group, Varian, and V.P.L. Research. As time goes on there could be more, depending on circumstances and public support, especially if it's Federal. Otherwise, meetings, hearings, proposals, and some testing will occupy the headlines until some group comes forward with objective results instead of small talk.

One trend does seem to be evident by both U.S. and current Euro-companies such as Philips and Tomson who are manufacturing receivers right now in this country. They do not want the 1125/60 Japanese standard, but the 1050/59.94 line and field rate already proposed by NBC and others who feel strongly that this is the way America should go, since it is easy to double the line rate but considerably more difficult to interfere with the field rate under which NTSC already operates. Should this come to pass, the Japanese, their video tape recorders, cameras, and other transmit/receive gear will certainly not be amused. It will be interesting to see how the FCC handles this paramount problem when it decides on an HDTV format in late 1991 or 1992.

High definition television can certainly be broadcast in the meantime over satellite and probably certain cable systems, and undoubtedly original MUSE will make up part of these experiments. But what satellite system finally triumphs will certainly require a working relationship with terrestrial HDTV, therefore 1050 lines of higher definition and resolution could well carry the day. As for digitized audio, Dolby seems now to have a complete system ready to go into production, and has for the moment a strong upper hand. When HDTV becomes a certainty, however, there could be other contenders for attentive ears.

Meanwhile, let's continue with a description of the various interests and contenders positions, as well as Congressional preoccupation with the subject, it is decidedly a political hot dog that needs both sustenance and rejuvenation.

ON CAPITOL HILL

On Pearl Harbor Day, December 7, 1988—47 years advanced from that "day of infamy," as Franklin Roosevelt named it—Chairman Edward J. Markey of the Telecommunications and Finance Subcommittee of the U.S. House of Representatives requested a broadened industry report "concerning methods for optimizing American participation in the development of advanced television technologies and derivative products." To be delivered by February 1, 1989, Congressman Markey wanted that report to offer a "blueprint for action" while Congress and the Executive Branch of government study "ramifications" of the high definition television revolution.

Originally proposed September 7, three months earlier, Mr. Markey asks if there is now need for Federal involvement, or should the outcome be left to private industry? If so, he asked, what should be the Congress, FCC, NTIA and any other Federal agency's involvement? And should the government set an HDTV standard for all the media, including broadcast, cable DBS, and VCR, as well as an HDTV standard for terrestrial broadcasters only or leave this decision to the marketplace? Should respondents answer "yes," Mr. Markey would want a criteria for such a standard and its timetable for development.

Next, the Subcommittee Chairman requested a market assessment for HDTV, along with a breakdown of potential for consumer electronics manufacturing, and related computer and semiconductor possibilities. He also wanted to know the economic implications of a U.S. system versus one of foreign origin, plus the possibilities of any antitrust "encouragement" exemptions, a matching grant program, or outright aid for an industry consortium such as solid state's SEMATECH to develop HDTV and related technologies.

OVER 24 RESPONSES

On the appointed day of February 1, 1989, American and European replies were promptly forthcoming. But Chairman Markey also heard from labor unions, universities, large electronics associations, HDTV proponents, and even a fellow Congressman, this time from California, who is most unhappy with the State Department. It all makes quite a report that we will cover in some detail, and should be somewhat of a definitive forecast for oncoming HDTV or EDTV proceedings. Eventual accomplishments or outright success, however, will have to await the 1990's transcriptions of history.

NYIT

The New York Institute of Technology, using a U.S. Dept. of Commerce report, predicts a $20 billion/year HDTV industry by the late 1990s and with 25% of U.S. homes owning at least one HDTV receiver by the end of this century. The effects, say William and Karen Glenn, will ultimately influence America's role "as a participant in worldwide trading markets." Included would be semiconductors, computers, receivers, etc.

The key to dominating the HDTV market, they say, will be the development of large, bright, and inexpensive means of displaying HDTV images. They point out that new technologies for this purpose are already being researched in Germany, Holland, and Japan. Such techniques, they continue, could "dramatically lower the cost and improve performance of HDTV receivers."

The Science and Technology Research Center of NYIT is also working on such development, and approximately half the Department of Defense initiative of some $14 million has been "earmarked" for such a project. But, the Glenns say, a considerably larger capital investment will be needed, probably on the order of $1 billion. And they warn that foreign governments and companies are already showing great interest in the solid state light valve projector (U.S. patent 3,882,271) that possesses "technical properties" to produce high quality HDTV displays. There is also a compact and inexpensive flat panel, fiberoptic screen device (U.S. Patent 4,116,739) that has prototypes already fabricated and now licensed by FiberView, Boulder, Colorado, that manufactures large fiberoptic screens for video stadium displays.

U.S. companies, they claim, must be "responsible for the design, development and manufacture of HDTV receivers if the U.S. is to realize the economic benefits of this new industry." And unlike the Japanese, the U.S. can innovate but has problems translating new ideas into marketable products, due basically, to short-term emphasis on immediate profit making. The Glenns would have a joint government/industry/academia consortium, similar to Sematech, undertake long-range projects of some 5-10 years for HDTV, in addition to a change in Federal tax structure for long range investment and capital gains.

AEA

The American Electronics Association with its 3700 members, as if to reinforce this outlook, also issued a lengthy reply to Chairman Markey citing U.S. strengths as a large homogeneous market, plenty of software, a considerable computer and telecommunications industry, a "healthier" but vulnerable semiconductor industry, existing AVT technology to develop, and future digital technology. Including such leaders in industry as Hewlett-Packard, Tektronix, AT&T, Apple-Computer, Cohu, T.I., National Semiconductor, Sun Microsystems, Digital Equip. Corporation, Ovonic Imaging, IBM, Intel, Motorola, Zenith, Raychem, Grass Valley Group, Varian, V.P.L. Research, and Anadigics.

Unfortunately, the United States lacks a viable, broad-based consumer electronics industry, and adequate levels of R&D, engineering, and manufacturing skills in consumer electronics to protect our market against outside domination. Incentives are needed for capital investment and technology in the consumer market and further access for the domestic semiconductor manufacturers to markets abroad.

President and CEO J. Richard Iverson adds that "AEA is fully committed to help organize and support formation of an industry-wide ATV strategy to build upon U.S. strengths." Some 20 AEA companies are now pledged to underwrite a business plan as the next crucial step in reviving this strategic, U.S.-based industry. Current market estimates are enormous, with current estimates between $40 and $150 billion. Failure to participate in ATV "will make U.S. national security dependent upon the availability of technological capability controlled by other nations." Without our own thriving industry, an outside standard is tempting, but AEA says the stakes are too high, and telecommunications, computers, and defense "are inextricably linked," and we cannot become dependent on another nation for our source of supply.

COMPACT

The Committee to Preserve American Color Television is another large organization, primarily labor unions, that has been active in trade policy matters, especially antidumping campaigns against TV makers from Japan, Korea, and Taiwan, and color picture tubes from Japan, Korea, Singapore, and Canada. Its statement says that antidumping orders now on the books, if properly enforced, will allow the U.S. a domestic market relatively free from unfair import competition. Further, COMPACT warns that U.S. color TV history "teaches" Japan's strong technological lead in HDTV will quickly be transformed into dominance of U.S. HDTV if strong government policies aren't soon adopted. Investment in a U.S. HDTV industry is needed, members say, for research and development in basic manufacture of "materials, components, and subassemblies, as well as final assembly of finished products. HDTV is an evolutionary product that should be produced by the domestic color television industry." COMPACT also attacked our government's handling of antidumping proceedings, saying the "contrast between the actions of our government and the activities of Japan's Ministry of International Trade and Industry was startling. Officials of our Commerce, Treasury, and Justice Departments who negotiated the settlement agreements . . . operated under a cloak of secrecy and did not even bother to consult with U.S. industry or its workers." COMPACT, however, says there is some hope because of antidumping orders on color CRTs in early 1988, and that perhaps the "downward slide" in the U.S. integrated producers financial performance will be reversed.

BELLCORE

The research and technical support arm of telephone companies Ameritech, Bell Atlantic, Bell South, NYNEX, Pacific Telesis, Southwestern Bell, and U.S. WEST—affirms that HDTV "represents a revolutionary improvement in visual communication." And Bellcore has already stocked laboratories with cameras, tape recorders, monitors, and switching equipment "to study and explain this new technology to national leaders, technologists and the citizenry." Bellcore sees the coming decade in terms of "information networking"—defined as access to information on demand from anywhere. HDTV is seen as one of its services available over fiber-based distribution structures that

214

will offer video service selection in "an unprecedented manner," including such offerings as football with subpictures on-screen of the quarterback, wide receivers, defensive backs, and whatever closeups you may wish in the overall picture.

Bellcore differs from many others by not wanting a single standard for the U.S., shortsighted, they say, because "it will close the door on innovation and prevent each of the media from realizing its full potential." Business users, they feel, will be singularly affected. Instead, Bellcore proposes a standard interface for all HDTV displays and calls for strong "government encouragement" for standards organizations. A view that most of Congressman Markey's respondents do not hold since they would largely prefer a single system for at least terrestrial HDTV, with perhaps some flexibility for cable satellite transmissions, and reception.

JAPANESE

The 1125/60 Hz Japanese "standard" accepted by Canada and originally by our own State Department is already under fire from five professors of the Massachusetts Institute of Technology with backing from Representative Mel Levine (D-CA). Congressman Levine questions if State's and the FCC's recent positions are compatible; if not, why not, and why the State Department should involve itself in technical matters.

Mr. Levine also states that the "United States has no choice but to develop its own HDTV industry if it is to remain on the industrial cutting edge . . . we will not remain a world leader in giant industries such as computers and automobiles—which will see 30% of their cost comprised of electronics by the year 2000—if we are not competitive in HDTV development . . . we must develop a comprehensive national strategy, led by the private sector, but supported by government, to develop it." He further stressed that Congress should move immediately to educate members about the importance of HDTV and "signal" the private sector we will support their efforts, and set up mechanisms to coordinate support for private sector HDTV development.

MIT

The Associate Director of MIT's Media Laboratory, Andrew Lippman, has other ideas than immediate HDTV development. He advocates forgetting television sets. In three years, he says, there won't be any. There'll be computers with high quality display screens and digital programming to receive "ABC, NBC, HBO, BBC, and anything we can dream up." Mr. Lippman would have us take our time, study the new wave of forthcoming computer appliances, and develop an open architecture receiver for universal formats. Taken to the extreme, he continued, there's no reason why a single channel shouldn't produce an HD sporting event, and three different but simultaneous newscasts, with fidelity possibly determined at each instant.

There's no need to rush, he argues, the U.S. doesn't have to buy anything we're not ready for because others have to negotiate with us. "We bring the world's largest, single, and most cohesive market to the table; what do they bring?" Other colleagues at MIT echo much of Mr. Lippman's reasoning, saying that any attempt to sell a new type of TV receiver with "somewhat sharper and wider pictures at a $10 \times$ cost over present sets would likely result in a marketing disaster." They, too, would emphasize digital processing, especially that of a computer inside a television receiver that "can be used for a greater diversity of programming choices within a given channel." And they don't want a single U.S. standard at this time, especially asking that the State Department quit supporting the Japanese-developed 1125/60/2:1 (line/Hz/interlace) format as the global HDTV standard since, they say, it will *not* become the U.S. standard. Respondents were W. Russell Neuman, Suzanne C. Neil, Lee McKnight, and Shawn O'Donnell of the MIT Media Lab. European manufacturers in the U.S., however, don't see eye-to-eye with MIT and want action now!

NAP

North American Philips, which manufactures Magnavox, Philco, and Sylvania brand receivers as well as Philips in Tennessee, has a $3 billion technology, research, development, and capital investment in the U.S. and thinks that competition is an "economy-wide issue," not limited to just HDTV.

NAP believes responsible government agencies should approve an NTSC-compatible terrestrial broadcast HDTV standard for 1050/59.94 line and field rate, applicable to broadcast, cable, and satellites. But HDTV's impact, however, should be viewed realistically, thinks Philips, and basic R&D plus manufacturing will offer more than simple competition and patent rights. Philips asserts that foreign-owned, but U.S.-based electronics companies can make vital contributions to HDTV with investments, expertise, and capabilities. And NAP would have the Federal government improve competitiveness by deficit reduction, direct more spending toward technology and education, encouraging industry tax policies, funding for research in semiconductors, displays, and manufacturing technology, amend antitrust laws for more cooperative efforts, and use international trade laws to open international markets and fair prices here at home.

Philips would also have the State Department cease supporting the 1125/60 proposal before the Consultative Committee on International Radio (CCIR) and back the U.S. 1050/59.94 proposed standard. There is no longer, they say, any possibility of a worldwide production standard. And company officials want a consumer based, NTSC compatible system with 16:9 aspect ratio, 4-channel sound, no artifacts, high definition resolution, better color, and pan-and-scan. Further, they want no delay in the decision-making process. Should HDTV be halted for all transmissions, "U.S. consumers will be left behind those of other nations in the ability to receive HDTV.

TCE

Thomson Consumer Electronics (RCA/GE), the David Sarnoff Research Center, and the National Broadcasting Company (NBC) responded to Chairman Markey in a joint reply, observing that Thomson is a global "participant" in the TV industry and "the largest television receiver manufacturer in the United States." Thomson's parent company has headquarters in France.

Like Philips, they say that future success for HDTV depends on planned development, and they want an evolving technology and NTSC compatibility with uniform standards for all. A phased-in approach, they continue, would offer a smooth path to a single ATV standard and would stimulate startup and market growth for both consumers and suppliers.

Initially a single 6 MHz NTSC-compatible system is needed, they suggest, with future additional spectrum for wider bandwidths, and they decidedly desire Executive Branch coordination of affected government agencies, especially the State Department.

Their Vice President of Thomson in Indianapolis, Indiana, declared that the real HDTV issue is jobs! Noting that NBC and Thomson are actively involved with Princeton's David Sarnoff Research Center in development of Advanced Compatible Television (ACTV), where some 35 years ago the Sarnoff Laboratories, NBC, and RCA Consumer Products developed and introduced NTSC color television to the U.S. and other nations. Dr. D. Joseph Donahue reports that, contrary to "widely held myths," R&D in the U.S. is not virtually non-existent and the three prime TV participants in this country, which are Thomson, Zenith, and Philips, represent about 50% of all color TV sales in America—all operating at U.S.-based facilities. "For the first time," Dr. Donahue reported, "American consumers will purchase more than 20 million color receivers this year . . . and, contrary to conventional wisdom, more than 6 out of every 10 of these color sets will have been produced or assembled in the U.S. by American workers. In fact, in the 25-inch and larger receivers, some 95% of the production and assembly is now concentrated in the U.S."

But he says that "given the global nature of the consumer electronics business today . . . it is . . . unrealistic to propose the creation of a strong American-owned consumer electronics industry." The real issue, according to Dr. Donahue, is "how to create the maximum number of American jobs in HDTV, R&D, and manufacturing, and how to ensure the maximum amount of U.S. content in HDTV products."

ZENITH

Zenith, the only remaining U.S.-owned and operated TV manufacturer, who keeps solvent by marketing its own brand products and private-labeling many more for such retailers as Sears-Roebuck and Montgomery Ward, says it would "take years," many currency devaluations, and aggressive trade policies plus enforcement, to even begin to correct the various trends that caused the loss of radio, stereo, and monochrome TV products, as well as component industries such as resistors, capacitors, ferrites, coils, etc. This loss, said Jerry Pearlman, Chairman and President of Zenith Electronics Corp., has caused a major technology drain.

As HDTV develops, Mr. Pearlman estimates that visual displays will amount to 50% and semiconductors 30% of receiver contents compared to 25% and 12%, respectively, for large screen sets today. And unless there is strong government action "to strengthen American technology," consumer, military, and industrial electronics will become even more dependent on foreign sources and technologies. Any action plan for HDTV, he continued, must have three main considerations:

1. Optimize standards for U.S. markets and keep patents and royalties in American hands.
2. Aid U.S. industry with funding for HDTV development, which includes both design and manufacturing dollars.
3. Restore domestic profitability with both government-supported R&D (just like Japan and Europe) as well as backing on international trade, especially antidumping laws. Conceding that it's "probably too late to recapture an American lead in passive component technology," Mr. Pearlman emphasizes it is not too late to save U.S. image displays and the semiconductor industries.

On another issue, Zenith sharply disagreed with the Electronics Industries Association on foreign versus domestic ownership. "The importance of domestic ownership of HDTV transmission system technology is far deeper than merely the assurance of manufacturing or R&D jobs in this country." Licensing royalties, Zenith suggested, "could amount to hundreds of millions of dollars each year by the turn of the century if current market projections for HDTV are met."

EIA

The Electronics Industries Association, for its part, has issued an excellent 98-page booklet on HDTV and Competitiveness of the U.S. Economy compiled by its own Advanced Television committee. "Competitiveness is a national concept," according to EIA, and "domestic firms can adopt production, investment, location and source strategies that weaken national competitiveness," but foreign-owned firms in the U.S. can operate with comparable strategies that will actually strengthen U.S. competitive conditions.

EIA says that to restore national competitiveness seven policy initiatives are recommended:

1. Reduce the deficit and increase national savings.
2. Shift Federal spending toward education, training, science and technology.

3. Change tax laws to encourage investment, especially investment tax credits and graduated capital gains.
4. Institute a permanent R&D development tax credit.
5. Use government money to encourage public-private enterprises for "middle-ground or generic technology projects," similar to Sematech.
6. Relax existing antitrust restrictions on cooperative R&D among private companies.
7. We need greater "openness" in international markets.

Today, it's said, that Tomson (RCA), Zenith, Philips (Magnavox, Philco, and Sylvania) are the three biggies in the TV business and own approximately 50% of the market. Further, the domestic parts content of U.S.-manufactured TV receivers is around 70% (possibly more). In the future, a study by Robt. Nathan Associates predicts that 13 million HDTV receivers will be sold in the U.S. in 2003, and 92% will be made in the U.S., and that HDTV will have 30% of the TV market by that year, contributing $23 billion to the national gross product and require 232,000 workers.

The study assumes FCC adoption of an HDTV standard by 1991; apparently a consensus among the better informed, but possibly not allowing for the standards squabble and further system testing and hardware development. In addition, we don't know if the FCC will extend the testing and design time should unforeseen circumstances arise, but they probably will.

As to an open architecture receiver, EIA says "current manufacturers of TV receivers are skeptical of this proposal because of increased costs and the creation of confusion among consumers." EIA, however, does favor a "friendly" multiport receiver with separate baseband and RF inputs, which probably also means RGB, YC, and possibly Teletext. The primary HDTV development and distribution problems as seen by EIA consist of HDTV-related ICs, large visual displays, new manufacturing technologies, and broadband switching. In NTSC sets of today, only 7-11% of the cost involves semiconductors, while in HDTV there should be about $400 worth. The big cost, at least in the beginning, will be for the large 16:9 cathode ray tubes.

CBS

CBS warns that until HDTV production and transmission systems have been developed, quality and diversified programming won't be available. The broadcasting network also wants a worldwide HDTV production standard primarily because of domestic program expenditures and international video software. "And should multiple transmission systems be used by the mass media," the CBS report said, "the market for an HDTV receiver capable of receiving any one of those systems will be smaller." On the other hand, CBS believes that multiple transmission systems will only shrink the HDTV receiver market and make sets more costly to manufacture. Finally, the broadcaster believes a competitive marketplace in HDTV manufacturing "is critical to the development of low-cost HDTV receivers." But CBS does not think that foreign manufacturers will assume a "dominant position" in HDTV and adversely affect American trade, jobs, and component industries.

SIA

SIA, the Semiconductor Industry Association, believes the Federal government should help optimize U.S. participation in HDTV technologies and "derivative products." This, the organization claims, "ultimately affects the entire U.S. electronics industry and other related industries and thereby has consequences for our nation's continued technological competitiveness, economic prosperity and economic security." SIA, in short, wants the reentry of U.S.-owned firms into high volume consumer products and production. A viable HDTV industry will depend, SIA says, on a healthy, competitive U.S. vendor base. Meanwhile, "measures" are being developed by SIA and AEA for vertical integration of U.S. electronics so they can compete directly with giant vertically integrated electronics companies in Europe and Asia.

AMST

The Association of Maximum Service Telecasters says that broadcasters are actively working to meet the HDTV challenge by helping to establish the Advanced Television Test Center, active in FCC Advisory Committee groups, and continue to educate broadcasters to prepare for the "coming revolution in TV technology." Should the U.S. fail to take firm action, AMST warns "it will by default permit the communications policies and priorities of other countries to determine the future shape of our domestic communications industry."

The telecasters would have Congress fund an ATV task force to aid the FCC in giving HDTV development the needed technical support; permit adequate spectrum for every local station to offer HDTV service; a single standard should be adopted for terrestrial transmissions; and Government should "ensure an advanced television system is developed to meet the needs of American broadcasters." AMST points out that more than 80% of TV programming is developed by local broadcasters.

NAB

The National Association of Broadcasters welcomes the time when viewers can receive advanced television service from their local stations. In general TV interests, however, NAB would urge the following:

- That Congress ensure the FCC takes no action prejudicial to the broadcast industry in its ability to compete with other ATV "distribution media."
- Establish a public Congressional policy to encourage free, over-the-air TV in an ATV environment.

■ Congress should back a single compatible ATV standard and so instruct the FCC.

■ And encourage U.S. research and development of ATV and related technologies, and consider grants or allowances for cooperative research and U.S. system development.

NAB recognizes Congress' responsibility to the industry, but wants no broadcaster holdups or conflicts with competing "pay" media. "U.S. Broadcasting needs a single, government-endorsed competitive ATV transmission standard . . . and soon . . . and it needs adequate spectrum to broadcast this new standard of television picture quality to the American public."

NAPTS

The National Association of Public Television Stations, a private, non-profit organization whose members are "licensees of virtually all the nation's public television stations," would adopt any new technology that "enhances its services to the American people." NAPTS hopes that Congress will provide "active oversight" of the FCC's deliberations on advanced television to ensure that the public's substantial investment in public television is protected and that the public interest is served.

The Corporation for Public Broadcasting is also a non-profit corporation authorized by Congress in 1967 "to facilitate and promote the development of public radio and television services for the American people." CPB wants to be sure that consumer use and enjoyment of video programming has prime consideration along with policy issues involving manufacturing and economic forces. It also wants the FCC to develop a single set of "coherent, mandatory ATV transmission standards sufficiently similar for terrestrial broadcasting and all other video media using ATV technology." Spectrum allotment should be decided on real-world testing, Congress needs to maintain continuous oversight of FCC activities regarding ATV for optimum ATV transmission standards, and also be cognizant of the "potential downside costs to the Nation's industries, particularly public television, and to the American public, if ATV develops in a manner detrimental to terrestrial broadcasting."

ABC

Capital Cities/ABC, Inc. wants HDTV technology available "within" the existing television system, and that government policy permit smooth transition to the new medium. Toward that end, these respondents want a single HDTV transmission standard since multiple standards would increase costs and result in consumer confusion. But in their view, "the promulgation of a worldwide production standard before the adoption of a U.S. transmission standard would not only be premature but could result in extreme disadvantage to our nation's interests." Capital Cities/ABC, consequently, is now appealing the decision of the American National Standards Institute (ANSI) to "accord national standard status to the 1125/60 proposal." Supporting letters have been received from the David Sarnoff Research Center, Zenith, Fox, Inc., Faroudja Laboratories, and the director of Advanced Television Research Program at MIT. Capital Cities/ABC also oppose any spectrum decisions before propagation tests have been analyzed and completed. In the meantime, the FCC should develop "a factual record on the spectrum needs for ATV and reach a policy judgment on what spectrum should be made available." In addition, they say, NTSC should be harmonically related to HDTV since non-harmonic line rates require interpolation to synthesize missing picture elements and lines, while non-integral frame or field rate relationships require either throwing away excess or synthesizing shortfall fields or frames. The first can lose resolution, particularly with motion, and the latter will result in periodic motion discontinuities. And it may not be possible to make NTSC and ATV downconversions acceptable to trained viewers, or program producers and directors. In short, they say, "conversion artifacts in the temporal domain are a key stumbling block to practical problems of using a world ATV standard as a source of program material for existing transmission standards and get all the proposed ATV systems."

DEL REY

The Del Rey Group, an associate of three major broadcasters in the Compatible Video Consortium, says its reply represents Del Rey only since there's not time to circulate it to the others. The Group was initially established to "introduce a compatible single-channel HDTV transmission system into North America." At the moment, they are defining technology and circuitry such receivers will eventually use, and "we are definitely not Japan bashing." This group would welcome governmental assistance as an "incubator" of small projects, but also look upon government as already being too large and business-intrusive. But many, such as this group needs help, too, and is shocked at the thought of the U.S. aiding a foreign-owned firm at the expense of domestic business, and questions whether European companies who have bought out U.S. consumer product companies "be let back in the game under special terms?" They endorse the idea of a consortium, especially if there is aid in funding.

OTHERS

The GTE Service Corporation believes in existing TV broadcast allocations; ATV should not be allowed to "undermine" the enormous investment in existing NTSC receivers; further information on ATV systems such as bandwidth requirements and certain other technical characteristics is required. GTE is currently building coaxial broadband facilities in Cerritos, California and wants a fiber facilities test bed with which to compare copper wire, coaxial and fiber-optics as alternative transmissions for voice, data, and video. GTE agrees with broadcasters who say it's too early to know if 6 MHz can offer ATV that will compete with HDTV delivered by other media. And how about cable handling passbands greater than 6 MHz? Some CATV installations are already at channel capacity, and wideband ATV would mean fewer channels/system. GTE would aid and cooperate with those who want to fully analyze technical, operational, and marketing issues challenging ATV deployment. GTE does not want to rush into sudden judgment and obviously feels that additional testing is necessary.

1125 GROUP

HDTV 1125/60 Group consists primarily of Japanese members and a few U.S. advisors who want the U.S. to support this production standard—the same as in Japan—and say that it should be regarded separately from transmission considerations. This, they claim, is the fastest and most efficient way to market HDTV for the benefit of many sectors of the economy, especially Hollywood and the billion dollar film export industry. And they observe that ATSC, SMPTE and ANSI all support the 1125/60 line/Hz production standard, they say that this standard is the only system specifically designed and easily convertible to all existing and proposed domestic and foreign transmission systems as well as 35mm film.

Faroudja Laboratories wants only a 1050/59.94 line/Hz standard that is both compatible and "friendly" to NTSC, or a 525 line with progressive scanning. Faroudja believes a single government body should be assigned "to take care of this project." Antitrust exemptions should be offered, along with new patent protection that will encourage the exchange of information between competing companies. Matching grant research funds are also sought, but only for new and improving technology, and a consortium established, patterned on MIITI of Japan that can become a "catalyst for future paths that American industry is to follow."

CONCLUSIONS OR CONTUSIONS

At this formulative juncture it seems that many have come "but few are chosen." Another semiannual meeting of Working Party 1 was held on May 8, 1989, and others will follow at approximately 6-month intervals until proponent positions are pretty well solidified and the learning curve among them has increased and possibly converged. If there are mergers and tradeoffs so that system numbers reduce from a dozen to just several, then both industry and the FCC will have a much easier task in selecting one or more to represent North American interests. And since Canada is working continuously with our own ATSC, that country presumably will go along with any reasonable findings. In the meantime, the various working parties and allied committees will keep digging for even more concrete results than have yet appeared.

By the latter part of 1990 we should also know something about test results, system survivors, and whether U.S. electronic associations will allow foreign-owned companies such as Tomson and Philips to have a piece of the action. So far, the tendency seems to remain all-American. We will also know if our own State Department and other governments at Geneva, Switzerland will actually approve the contested 1125/60 lines/fields as a world standard. Advocating and signed treaties are two different bit streams that require parallel sampling and decoding for positive verification. Geneva subcommittees and our own can recommend, but the world body of nations controlling the electromagnetic energy field have the final say. Whether yes or no, it is rather apparent now that European and U.S. requirements are different and could well remain separate for domestic consumption. And whether transcoding the two for full fidelity can become a reality, no one positively knows just yet.

Nonetheless, the die is cast, there will most certainly be HDTV in Japan, Europe, Canada, the U.S., France, and elsewhere, but exactly when becomes a matter of interesting conjecture. Some predictions, in the form of charts, can offer a reasonable idea of what, if we only knew when . . . Therefore, all are predicated on Year No. 1, when HDTV becomes available. For the remainder, your guess is just as good as mine, possibly better if you're especially lucky. You now know all that I know, and that's that!

PENETRATION OF
U.S. HOUSEHOLDS (%)

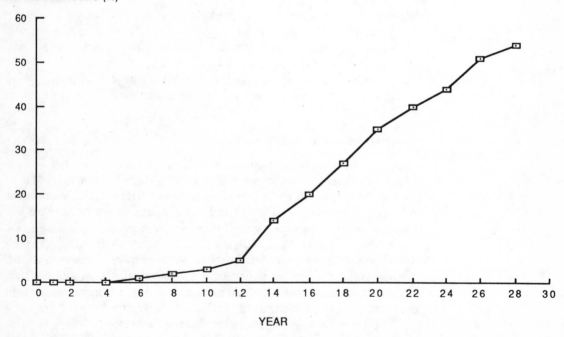

YEAR

Note: Year 0 is the year of ATV receiver marketplace introduction. One percent household penetration is achieved around Year 8.

Fig. 10-1. WP-5 projection of ATV receiver penetration.

PENETRATION OF
U.S. HOUSEHOLDS (%)

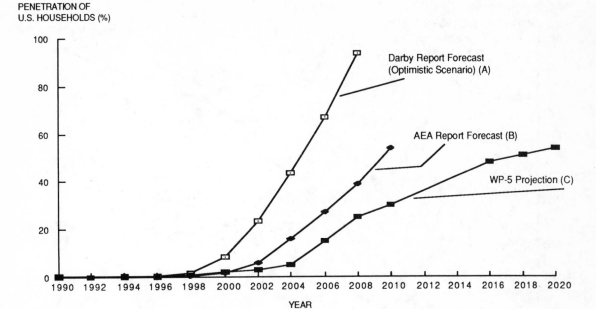

Notes: (A) Darby forecast has 1% household penetration achieved in 1997, 7 to 8 years after product introduction.

(B) AEA forecast has 1% household penetration achieved in 2000, 10 years after product introduction.

(C) WP-5 did not forecast year of introduction, but specifies that 1% household penetration will be achieved about 8 years after product introduction. 1990 was chosen as year of introduction in this Exhibit for purposes of comparison with the other forecasts.

Fig. 10-2. Comparison of ATV receiver penetration forecasts.

Planning Subcommittee Working Party 5

Three graphs have been offered by the Working Party in an attempt to predict HDTV receiver market penetration in the "Year 1" that begins several years in the future. Figure 10-1 is the WP-5 forecast; Fig. 10-2 is a combination of the Darby Report, another by the American Electronics Association, and WP-5; and Fig. 10-3 is a WP-5 VCR and color TV annual growth projection compared with what is expected of HDTV. Whether such remains or changes in the future, probably depends on HDTV receiver prices, publicity, the actual systems in use, and the various manufacturers offering them. Consequently, some curves could change radically with experience and time. But for right now, these are about the best we have.

The three drawings are published exactly as received from WP-5 without alteration.

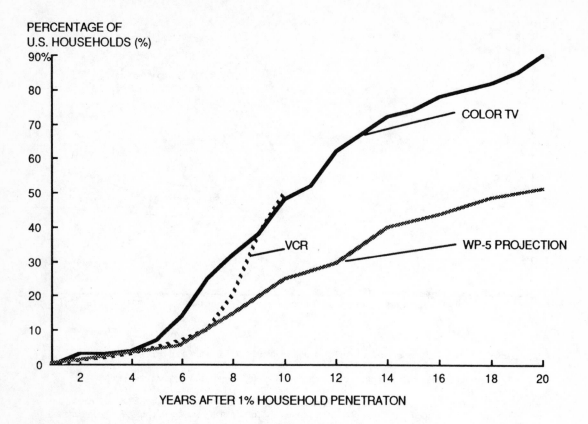

PERCENTAGE OF
U.S. HOUSEHOLDS (%)

YEARS AFTER 1% HOUSEHOLD PENETRATON

<u>Source for color TV and VCR data</u>: National Association of Broadcasters, "New Video Technologies:
Consumer Acceptance and Price Histories," November 10, 1987.

<u>Notes:</u>

(A)   NAB cites 1961 as the year when color TV reached 1% penetration of U.S. households

(B)   NAB cites 1978 as the year when VCR reached 1% penetration of U.S. households

Fig. 10-3. Penetration rates of color TV and VCR versus WP-5 projection of ATV receiver penetration (after 1% household penetration).

Glossary

artifacts: details displayed on the screen that do not accurately represent the actual scene.

aspect ratio: the ratio of picture width to height.

band-limiting: restricting a signal to a maximum bandwidth.

baseband: of or referring to the frequencies that contain all the modulating signals applied to a transmitter.

chrominance: the transmitted color information of a picture.

color difference signal: color information transmitted in the form of two signals.

compatibility: in the context of this proceeding, usually the ability of an NTSC receiver to receive and display an ATV signal with quality not significantly degraded from that displayed by reception of a conventional NTSC signal (see ¶80 of the *NOI*).

contiguous channel: the channel immediately adjacent in frequency to another channel.

cross color: artificially created color ''details'' of a picture caused by an unwanted interaction between signals containing the color information that results in inaccurate colors being created and displayed on the screen.

cross luminance: artificially created brightness information caused by an unwanted interaction between the color information and the brightness information. False brightness is thus created and displayed on the screen.

detail processors: devices that improve the displayed image.

digital audio: audio information that has been digitized for transmission; digitization is the conversion of an analog signal into a series of ones or zeros.

encoders and decoders: encoders convert a signal into another signal; decoders restore the original signal. These processes reduce the bandwidth required to transmit television signals.

flesh tone correctors: devices that restore the original colors of an image; the human eye is most sensitive to the color accuracy of skin tones.

frame rate: the rate at which entire pictures are transmitted.

frame store: electronic memory in a television receiver used to store an entire picture (called a frame).

frequency division multiplexing: the transmission of a single signal, with different information sent at different frequencies.

image: a scene portrayed on the television screen.

luminance: the brightness information of a television picture.

Multiple Analog Component (MAC) systems: systems that transmit a television signal by separating it into several components that are time diversity multiplexed together; to be distinguished from composite systems, such as NTSC, which combine the various signal components into a single signal by frequency division multiplexing.

progressive scanning: a system of displaying a picture in which the entire picture is shown in one continuous sequence; to be distinguished from interlaced scanning, in which the entire picture is displayed in two consecutive sequences or fields. Viewers must mentally integrate the two fields to perceive the picture.

panning and scanning: panning is the act of moving a camera slowly from one side of a scene toward the other; scanning captures the scene electronically, after panning towards the subject.

quadrature modulation: method of adding additional information to a signal by creating a new frequency component and delaying its transmission by a quarter of the time it takes to transmit one radio frequency cycle; quadrature modulation is used to add color information to the black and white NTSC signal.

resolution: a picture's fineness of detail; high-resolution information is needed to provide a very finely detailed picture.

retrace intervals: time period during which the horizontal scanning line returns to the far left side of the screen or to the top of the next field (horizontal and vertical retrace intervals).

sidepanels: detail additional to that transmitted in NTSC systems that is required for a picture with a wider aspect ratio.

signal processing: electronic enhancement of a transmitted signal performed by the receiver.

sound carrier: the reference frequency upon which the audio signal is impressed (modulated).

spatial resolution (also called static resolution): fineness of detail of a picture that is without motion.

subpixel: (utilized by the Del Rey Group HD-NTSC system) a subdivision of each picture element (pixel) into thirds; one subpixel is sent in each of three successive frames.

subsampling: a method of reducing the bandwidth required to transmit a signal.

sampling: describing a signal with another signal.

temporal resolution: an indication of the accuracy of motion portrayal.

three dimensional filtering: a picture is a time sequence of horizontal and vertical detail (a flat view of a scene changing in time); if three dimensional filtering is employed, the scene is simplified horizontally, vertically, and temporally.

time division multiplexing: transmission of several signals consecutively, according to a predetermined sequence.

transcoder: a device used to convert a signal with one scanning rate to another.

UHF taboos: a set of NTSC transmission prohibitions based upon certain UHF television channel combinations and transmitter mileage separations that the FCC determined are required to prevent interference to television receivers; most of the taboos are caused by receiver characteristics.

video carrier: reference frequency upon which video information is impressed (modulated). The resolution information is impressed upon the picture carrier and the color information is impressed upon the color subcarriers.

Index